Traces of a Mobile Field

This agenda-setting collection critically reflects upon a decade of contributions to the social scientific 'mobilities turn' in order to propose new trajectories for the future of this interdisciplinary research field. The chapters are all exemplars of how the past decade of research has opened up new insights into the place of mobilities in societies. They also highlight how attempts to look forward towards new conversations, understandings, and interventions in a mobile world will emerge from the transformations invoked by this field of research. Authors foreground issues of power, interdisciplinarity, transformative technologies, fragmented discourses and changing social processes whilst addressing automobility, aeromobility, tourism, communications technologies, urban infrastructures, migration, and emergencies. As a whole, the collection raises important questions about not only how understandings of mobilities are changing, but also how the field of mobilities research is itself on the move. The evocative empirical cases and provocative arguments in this book thus highlight the necessity of new concepts, conversations, methods, empirical studies and interventions to address transformations in both the complex mobilities of social worlds and what is examined or taken for granted in mobilities research itself. This book was originally published as a special issue of *Mobilities*.

James Faulconbridge is Professor in the Department of Organisation, Work and Technology at Lancaster University Management School, UK. His research focuses, in particular, upon the way forms of mobility are used in global firms, with the role of business travel being of especial interest.

Allison Hui is an Academic Fellow in Sociology and the DEMAND Centre at Lancaster University, UK. Her research examines transformations in everyday life in the context of changing global mobilities, focusing particularly on theorising social practices, consumption and travel.

Traces of a Mobile Field

Ten Years of Mobilities Research

Edited by
James Faulconbridge and Allison Hui

Routledge
Taylor & Francis Group
LONDON AND NEW YORK

First published 2017 by Routledge

2 Park Square, Milton Park, Abingdon, Oxfordshire OX14 4RN
52 Vanderbilt Avenue, New York, NY 10017

Routledge is an imprint of the Taylor & Francis Group, an informa business

First issued in paperback 2018

British Library Cataloguing in Publication Data
A catalogue record for this book is available from the British Library

ISBN 13: 978-1-138-70858-7 (hbk)
ISBN 13: 978-0-367-14242-1 (pbk)

Typeset in TimesNewRomanPS
by diacriTech, Chennai

Publisher's Note
The publisher accepts responsibility for any inconsistencies that may have arisen during the conversion of this book from journal articles to book chapters, namely the possible inclusion of journal terminology.

Contents

Citation Information vii
Notes on Contributors ix

1 Traces of a Mobile Field: Ten Years of Mobilities Research 1
James Faulconbridge and Allison Hui

2 Uneven Mobility Futures: A Foucauldian Approach 15
Mimi Sheller

3 Emergency Mobilities 32
Peter Adey

4 Re-Assembling (Aero)mobilities: Perspectives beyond the West 49
Weiqiang Lin

5 The Boundaries of Interdisciplinary Fields: Temporalities Shaping the Past and Future of Dialogue between Migration and Mobilities Research 66
Allison Hui

6 Mobility Infrastructures: Modern Visions, Affective Environments and the Problem of Car Parking 83
Peter Merriman

7 Mobilities and Urban Encounters in Public Places in the Age of Locative Media. Seams, Folds, and Encounters with 'Pseudonymous Strangers' 99
Christian Licoppe

8 Mediated Pedestrian Mobility: Walking and the Map App 117
Eric Laurier, Barry Brown and Moira McGregor

9 The Anticipated Futures of Space Tourism 135
Mark R. Johnson and Daryl Martin

10 Accelerate, Reverse, or Find the Off Ramp? Future Automobility in the Fragmented American Imagination 152
Katherine G. Reese

Index 171

Citation Information

The chapters in this book were originally published in *Mobilities*, volume 11, issue 1 (February 2016). When citing this material, please use the original page numbering for each article, as follows:

Chapter 1
Traces of a Mobile Field: Ten Years of Mobilities Research
James Faulconbridge and Allison Hui
Mobilities, volume 11, issue 1 (February 2016) pp. 1–14

Chapter 2
Uneven Mobility Futures: A Foucauldian Approach
Mimi Sheller
Mobilities, volume 11, issue 1 (February 2016) pp. 15–31

Chapter 3
Emergency Mobilities
Peter Adey
Mobilities, volume 11, issue 1 (February 2016) pp. 32–48

Chapter 4
Re-Assembling (Aero)mobilities: Perspectives beyond the West
Weiqiang Lin
Mobilities,volume 11, issue 1 (February 2016) pp. 49–65

Chapter 5
The Boundaries of Interdisciplinary Fields: Temporalities Shaping the Past and Future of Dialogue between Migration and Mobilities Research
Allison Hui
Mobilities, volume 11, issue 1 (February 2016) pp. 66–82

Chapter 6
Mobility Infrastructures: Modern Visions, Affective Environments and the Problem of Car Parking
Peter Merriman
Mobilities, volume 11, issue 1 (February 2016) pp. 83–98

Chapter 7
Mobilities and Urban Encounters in Public Places in the Age of Locative Media. Seams, Folds, and Encounters with 'Pseudonymous Strangers'
Christian Licoppe
Mobilities, volume 11, issue 1 (February 2016) pp. 99–116

Chapter 8
Mediated Pedestrian Mobility: Walking and the Map App
Eric Laurier, Barry Brown and Moira McGregor
Mobilities, volume 11, issue 1 (February 2016) pp. 117–134

Chapter 9
The Anticipated Futures of Space Tourism
Mark R. Johnson and Daryl Martin
Mobilities, volume 11, issue 1 (February 2016) pp. 135–151

Chapter 10
Accelerate, Reverse, or Find the Off Ramp? Future Automobility in the Fragmented American Imagination
Katherine G. Reese
Mobilities, volume 11, issue 1 (February 2016) pp. 152–170

Notes on Contributors

Peter Adey is a Professor at the Department of Geography, Royal Holloway University of London, Surrey, UK. His work lies at the intersection between space, security and mobility, and the blurring boundaries between cultural and political geography.

Barry Brown is co-director at the Mobile Life Centre, Kista, Sweden. His recent research has focused on the sociology and design of leisure technologies.

James Faulconbridge is Professor in the Department of Organisation, Work and Technology at Lancaster University Management School, UK. His research focuses, in particular, upon the way forms of mobility are used in global firms, with the role of business travel being of especial interest.

Allison Hui is an Academic Fellow in Sociology and the DEMAND Centre at Lancaster University, UK. Her research examines transformations in everyday life in the context of changing global mobilities, focusing particularly on theorising social practices, consumption and travel.

Mark R. Johnson is now a Postdoctoral Fellow at the Digital Creativity Labs, University of York, York, UK.

Eric Laurier is a Reader in Geography & Interaction at the School of GeoSciences, Institute of Geography, University of Edinburgh, Edinburgh, UK.

Christian Licoppe is Professor of Sociology and head of the Department of Social Science, Telecom Paristech, France. Among other things he has worked in the field of mobility and communication studies for several years.

Weiqiang Lin is now an Assistant Professor in Geography at the National University of Singapore. His research is primarily focused on aviation, mobilities and infrastructures.

Daryl Martin is a Lecturer in Sociology, Department of Sociology, Wentworth College, University of York, York, UK. His research interests lie in the areas of architectural theory, cultural geography, social theory and urban studies.

Moira McGregor is based at the Mobile Life Centre, Kista, Sweden. She has a keen interest in field research, ethnography and ethnomethodology in relation to HCI.

NOTES ON CONTRIBUTORS

Peter Merriman is Professor in Human Geography at the Department of Geography and Earth Sciences, Aberystwyth University, Aberystwyth, UK. His interdisciplinary research focuses on mobility, theories of space and place, architectural geographies, nationalism and national identity, and the cultural-historical geographies of 20th Century Wales. He is one of the world's leading scholars in mobility studies.

Katherine G. Reese is a doctoral student in the School of International Service, American University, Washington DC, USA. She also serves as managing editor for the *Journal of International Relations and Development.* Her interests revolve around international communication, mobility, and environmental politics.

Mimi Sheller is Professor of Sociology at the Department of Sociology, and Director of the Center for Mobility Research and Policy, College of Arts and Sciences, Drexel University, Philadelphia, USA.

Traces of a Mobile Field: Ten Years of Mobilities Research

JAMES FAULCONBRIDGE* & ALLISON HUI**

* *Department of Organisation, Work and Technology, Lancaster University Management School, Lancaster University, Lancaster, UK*

** *Department of Sociology, Lancaster University, Lancaster, UK*

ABSTRACT *Since the launch of this journal 10 years ago, the field of mobilities research has developed at a rapid pace. In this editorial introduction, we explore how this development has been curated, how the field has evolved and what maturation might mean for mobilities research. After reviewing how early editorials encouraged particular trajectories of development within mobilities research, we introduce the papers in this special issue, which build upon and re-shape key discussions that have emerged in the last decade. Drawing out issues of power, interdisciplinarity, social processes and futures, the papers raise important questions about not only how understandings of mobilities are changing, but also how the field of mobilities research is itself on the move. Taking up these themes, we examine how understanding mobilities research as a field, contributes to considerations of the potential for future struggles, fragmentation and sub-disciplines. We argue that the open nature and strategic diversity of the mobilities field has fed the successes of the past decade, and therefore needs to remain a priority in the future – with a careful balance curated between convergence around key themes and the exploration of varied 'internal goods' which remain an important source of inspiration and creative potential within the field.*

Introduction

In September 2013, the paths of many colleagues and friends coalesced at the 'Global Mobility Futures' conference at Lancaster University, UK, to celebrate the tenth anniversary of the Centre for Mobilities Research (CeMoRe). This gathering sparked many exciting conversations about mobilities research, a field that some of the participants first explored during CeMoRe's 2004 conference on 'Alternative Mobility Futures'. More importantly, however, it affirmed that interactions within mobilities research remain vibrant, creative and generative. In this special issue, which emerged from the 2013 conference and also marks 10 years since the launch of this journal in 2006, we seek to capture the creative momentum of mobilities research and critically reflect upon how the field emerged and what its future might hold.

As a journal, *Mobilities* was born at a crucial moment. Many interdisciplinary discussions about the topic of mobility were emerging, in eclectic and dispersed contexts. By bringing together ongoing conversations, and providing a forum for further dialogue, the journal helped to encourage a convergence of interests around the dynamics, stakes and empirical study of mobilities. Alongside the first issue of the journal, the year 2006 also saw the publication of a special issue in Environment and Planning A. The editorials for the two issues (Hannam, Sheller, and Urry 2006; Sheller and Urry 2006) have since become the starting point for most discussions of the 'mobilities turn'. From this point on, the last decade has seen a proliferation of mobilities scholarship, as testified by the success of this journal (which has expanded from three to five issues per year) and the many published handbooks (Adey et al. 2014b), special issues (Anim-Addo, Hasty, and Peters 2014; Bajc, Coleman, and Eade 2007; Barker et al. 2009; Basu and Coleman 2008; Conradson and Mckay 2007; Fortier and Lewis 2006; Gill, Caletrío, and Mason 2011; Richardson 2013) and reviews of the field (Cresswell 2011, 2012, 2014; Merriman 2015).

It could be tempting on an anniversary such as this to summarize and to ask what is new about the scholarship published under the badge of mobilities research. Inevitably, though, answering such a question is an impossible task. Not only do assessments of novelty vary in relation to how they are situated and contextualized, presenting particular problems for interdisciplinary fields, but such assessments would also be slippery and fleeting given the heterogeneity and continual development of the field.

Indeed, the diversity of the papers in this special issue is a testament to both the dynamic development of mobilities research and its broad scope. After 10 years, the vibrant and interdisciplinary set of dialogues that characterize the field has largely rendered moot the question of whether there really was a mobilities turn or paradigm shift. More interesting, therefore, are questions of how links and connections, histories and futures, become intertwined in 'state-of-the-art' mobilities research. The papers in this special issue are all exemplars of how the past decade of research has opened up new insights into the place of mobilities in societies. They also highlight how attempts to look forward towards new conversations, understandings and interventions in a mobile world will be embedded in, and emerge from, the more or less deliberate transformations invoked by the rise of the mobilities field. They remind us, then, that the state of the art has its own history, politics and trajectory.

Therefore, in this editorial, rather than attempting to summarize the diverse empirical, theoretical, and methodological themes of this special issue, we step back to critically reflect upon mobilities as a field of study with its own past and future. This stems from the conviction that the vitality of mobilities research is tied to not only the processes it studies, but also the practices of mobilities researchers. Ensuring the vitality and creativity of mobilities research in the next decade is, therefore, not only about reflecting upon the understandings and politics of the mobilities we study, but also about situating our own interactions within the understandings and politics of research fields.

Our reflections begin by considering how, in 2006, the future of the nascent field of mobilities research was shaped by editorial calls to join an interdisciplinary and empirically driven project. We argue that these initial framings of mobilities research were central in shaping what was to become a strategically diverse field. After introducing how the special issue papers represent traces of the field which gather and redirect existing conversations, the second half of the paper considers more closely what it means to address mobilities research as a 'field' (Bourdieu 1984). Engaging with discussions of the emergence and maturation of organizational fields, as well as the struggles and rewards of competition within fields, it highlights how treating mobilities research as a field has implications for imagining the next decade of development.

Curating New Trajectories

Our initial point of departure is to reflect upon the genealogy of mobilities research and how particular trajectories within it have enabled creativity and contributions in the last decade. While authors in Adey et al.'s (2014b) recent handbook consider the genealogies of several (sub)disciplines contributing to work on mobilities, we turn instead to the two 2006 editorials and how they both recognized and encouraged particular lines of exploration. These initial interventions, we suggest, have significantly energized subsequent work and contributed to the formation of a distinct, interdisciplinary field of mobilities research.

Though the editorials by Hannam, Sheller, and Urry (2006) and Sheller and Urry (2006) are regularly cited in reference to a 'mobilities turn' or 'mobilities paradigm', it is important to recognize that their contribution was largely one of curation and categorization. The imperative of both was to make a discursive injunction by trying to name and re-categorize work in order to develop a mobilities agenda. The authors were thus careful to position the mobilities turn as something that in many ways had already happened. Subsequent tensions and discussions have arisen around this labelling – how for instance it relates to the historicization of mobilities scholarship (Cresswell 2010) and the existence of coherent paradigms within the social sciences. This can be seen as evidence of how this knowingly 'provocative labelling' (Adey et al. 2014a, 2) provided momentum for new discursive engagements and a re-thinking of previous segmentations and separations within social scientific literature. There is, nonetheless, still an outstanding question: what made the discursive, if not paradigmatic, shift signalled by the two editorials compelling, given that there were already decades of mobilities studies within other disciplines such as anthropology, geography, migration and transport studies?

In our reading, the editorials' key contribution was to propose a transformation and re-arrangement of research practices in order to foreground mobility, rather than seeing it as a constituent of larger social processes. This implies the need to investigate empirical cases involving mobilities without initially presuming that they can be explained through existing theoretical or disciplinary framings. In this sense, mobilities are not 'superfluous' or 'exceptional', but worthy of extended attention (Adey et al. 2014a, 3). Mobilities research involves a 'style of analysis' in which mobilities are treated critically and in context (Adey et al. 2014a, 2). In some ways, this move can be read as a simple validation of mobilities as objects of study. But it also created a space for considering what was, at the time, not orthodox in some disciplines such as sociology (Sheller 2014; Urry 2000). Encouraging efforts to put mobilities centre stage thus implicitly challenged various (sub)disciplinary assumptions of what was relevant and irrelevant, consequential and inconsequential, central and peripheral within the social world. The invitation to consider mobilities provided, then as now, a space to question 'normal' assumptions and categories, challenge segmentations and boundaries within communities of knowledge, and explore creative and experimental approaches to studying the social world.

This call to examine mobilities on its own terms suggests, but is not premised upon, particular theoretical and methodological approaches. Indeed, notable has been the resultant theoretical and methodological experimentation associated with the study of diverse empirical topics (as the papers in this special issue illustrate). While, for instance, Urry has outlined a set of theoretical concepts that support considerations of mobilities (2007, chapter 3), these do not represent an orthodoxy or construct a new grand narrative. Rather, looking back over the past decade of research, it is possible to identify streams of scholarship that develop different theories, methods and 'styles' of analysis in order to understand mobilities. This is reflective of the encouragement given in the two editorials to study diverse empirical cases and retain a sense of openness about the theoretical and methodological resources that might be appropriate. Hence, 'The new mobilities paradigm suggests a set of questions, theories, and methodologies rather than a 'totalizing' or reductive description of the contemporary world' (Sheller and Urry 2006, 210). This suggests, and research published in the last decade appears to confirm, a primary aim of creating a space where studies of diverse mobilities might be investigated without the constraints imposed by pre-existing disciplinary frameworks.

Though contemporary theoretical and methodological approaches remain as heterogeneous as those in the initial lists suggested by Sheller and Urry (2006), several important trajectories can nonetheless be identified. Theoretically, for instance, we note the rise of an ontological position that is more process-, performance- and practice-oriented, with mobilities being seen as emergent and in need of theorization as such. This involves, for instance, the use and development of work by Goffman (Jensen 2006, 2010; Licoppe 2009; Yeoh and Huang 2010) and de Certeau (Bissell 2009; Farías 2010; Kidder 2009), as well as on affect (Bissell 2010; Conradson and Latham 2007; Jensen, Sheller, and Wind 2014), practices (Aldred and Jungnickel 2013; Benson 2011; Cresswell and Merriman 2011; Hui 2013; Larsen 2008a), creating networks (Blok 2010; Hui 2012; Larsen 2008b; Larsen, Urry, and Axhausen 2006; Nowicka 2007; Ren 2011) and the non-representational (McHugh 2009; Spinney 2011; Vannini 2011). Relatedly, empirical analyses have privileged not the functionality of moving from A to B but experiences and sociocultural constructions of mobilities (Cresswell 2006; Jensen 2009). The methodological challenges and opportunities of studying mobilities have also contributed to the emergence of 'mobile methods' that combine and re-orientate traditions such as ethnography with the use of new technologies such as the smart phone (Buscher and Urry 2009; Büscher, Urry, and Witchger 2011; Fincham, McGuinness, and Murray 2010; Hein, Evans, and Jones 2008). All of these developments stem from prioritizing the specificities of mobility itself, and mark a contrast with studies in which mobility is treated as an incidental part of a wider phenomenon.

The development of particular theoretical and methodological approaches is also accompanied by a number of distinctive characterizations of the phenomena that fit under the mobilities umbrella. These are even more diverse than the theoretical and methodological developments outlined above. This was perhaps inevitable since the editorials a decade ago explicitly promoted interdisciplinary investigations of mobilities (Hannam, Sheller, and Urry 2006, 2, 16). An interest in creating dialogue across existing disciplinary boundaries not only suggested but required conceptual openness: 'Mobilities in this paradigm is thus used in a broad-ranging generic sense' (Sheller and Urry 2006, 212). Hence, different types of movement (of people, objects, capital, information), as well as their intersections and dynamics across multiple systems and scales, have been of concern (Hannam, Sheller, and Urry 2006, 1). An attention to connections and flows, exclusions and obstacles, has highlighted how heterogeneous mobilities reproduce systems of governance and politics, as well as situated experiences and engagement. All of this creates space to move beyond some of the more traditional foci for research, such as embodied travelling by car, train or any other means. Mobilities research has taken such cases seriously, but emphasized the importance of connecting them to broader processes and problems.

Since the 2006 editorials, mobilities research has thus emerged around both shared interests and important departures. These have contributed to many exciting trajectories of development that are addressed, in part, by the papers in this special issue.

Traces of a Mobile Field: The Special Issue Papers

The above description is, of course, no more than a snapshot of developments over the past decade. Likewise, the papers in this special issue gather a diverse set of traces of the ever-shifting field of mobilities research. They are, for one, traces of the anniversary conference and its heterogeneous conversations. But they also trace major areas to which mobilities research has contributed in the past decade, and to which it may contribute in the future. In this way, they represent an (inevitably) incomplete picture of advances and accomplishments that have emerged over the past decade, continually re-making understandings of 'state-of-the-art' mobilities research. Uniting all of these papers, as would be expected, is the style of interdisciplinary scholarship that has come to define the field in the past decade; analyses defamiliarizing apparently familiar topics and allowing mobility in and of itself to emerge centre stage. Hence, each paper forges links to varied histories of mobilities research, extending some trajectories and redirecting others. Histories are linked to patterns of inequalities, discourses of emergence and development, and varied interdisciplinary encounters. The papers also connect and move between mobile units, terrains of negotiation and contestation, and groups of differentially mobile subjects. In this way, they sit at the intersections of many still-developing trajectories of mobilities research.

The first two papers (Sheller, this issue; Adey, this issue) reflect upon evolving research agendas, asking to what conversations and problems mobilities research might better contribute. In particular, inequalities, highlighted by Kaufmann et al. (2004) as a constituent feature of mobility, and emergencies are highlighted as areas in need of further theoretical development and attention. Sheller engages with the work of Foucault, and Adey broader themes around governance, in order to build up theoretical resources for addressing how the unevenness and politics of mobilities are manifest in a range of sites and cases. Their discussions highlight how the lack of a grand theory within mobilities research need not lead to the ignorance of systemic processes and problems – rather, mobilities scholars build up appropriate interdisciplinary vocabularies and understandings to fit the complex and changing sites they study. These contributions thus suggest the potential for new agendas and theoretical development in the next decade of research.

Subsequent papers by Lin and Hui (this issue) consider how mobilities research might be further developed through engagement with cases that are exceptional within existing discussions. A range of studies on aeromobility over the past decade (Adey 2008; Budd 2011; Burrell 2008; Cwerner, Kesselring, and Urry 2009; Lassen 2006) inform the work of Lin, who highlights the Western biases of many existing studies and the corresponding failure to understand how significant aeromobile assemblages are for non-Western economies and societies. Hui reflects upon interdisciplinary dialogues between migration and mobilities scholars, unpicking key points of friction and making the 'migrant exceptionalism' perpetuated by both groups both a target of critique and an inspiration for future interdisciplinary engagement. In different ways, these papers show that reflecting upon how mobilities research is conducted might open up new empirical sites, new interdisciplinary interactions and new understandings of mobilities themselves.

Papers by Merriman and Licoppe (this issue) then examine relationships between complex transformations and mobile things in urban environments. Though Sheller and Urry noted a dearth of research on automobility in their editorial (2006), this has been redressed by mobilities scholars in the last 10 years (Böhm et al. 2006; Conley and McLaren 2009; Featherstone, Thrift, and Urry 2005). Such discussions are here extended by Merriman, who highlights how urban environments have been transformed by not only cars, but by the processes of infrastructuring that surround materialities and imaginations of car parking. Licoppe then considers a different aspect of urban transformation, reviewing how mobile locative media have transformed mobile subjects and encounters through the creation of 'pseudonymous strangers' and the negotiation of the seams and overlaps between virtual and physical spaces.

The embodied experience of mobile subjects has been another important area to which mobilities research has contributed in the last decade. Laurier et al.'s paper (this issue) takes up the intersection between mobile subjects and mobile methods, looking at how embodied, interpersonal performances of walking become reconfigured through the use of mobile devices and map apps. Drawing upon detailed conversation analysis of video and screen-captured data, their paper highlights how the exploration of new technologies in mobile methods have made it possible to attend closely to the meanings and movements of mobile bodies.

The final two papers by Johnson and Martin (this issue) and Reese (this issue) draw attention to the centrality of questions about futures in mobilities research. For Johnson and Martin, this relates to new technologies of mobility, in the form of space travel, and how actors within this emerging sector draw upon discourses of tourism and other modes of transportation to envision future mobilities. In contrast, Reese looks at an 'old' technology – the automobile – asking how the stories we tell about automobility have implications for the future, and in particular for reducing carbon emissions as an urgent public policy priority. These final papers remind us, then, both of the importance of discourses and imaginations of mobilities past, present and future, and of the intricate relationships between the field of mobilities research and the ever-changing societies that we study.

While these special issue papers highlight, in different ways, the trajectories of mobilities research introduced in the previous section, they also raise questions about what happens next and where mobilities research goes from here:

- How might mobilities research remain exciting and compelling, while taking seriously the power dynamics that mark both research practices and the cases they study?

- To what extent has mobilities research not only challenged some assumptions and boundaries, but also started to reproduce others?
- To what extent is coalescence around a number of theories and associated debates needed? What benefits or risks could this entail?
- How do the stories we tell about mobilities research suggest what is permanent or changeable about the practices it reproduces or studies in the future?

These questions are important because of what is at stake in the study of uneven mobilities and unsustainable mobilities. They are also important because, as the last decade has shown, how studies of mobilities are framed and approached has implications for the diversity of subsequent contributions. The following sections therefore draw upon understandings of 'fields' (Bourdieu 1984) to engage with these questions, reflect critically upon the last decade of mobilities research and imagine what the next decade might bring.

The Politics of Emerging Research Fields

The openness of 'mobilities' as a concept has undoubtedly enabled exciting and fruitful discussions that challenge 'normal' categories and units of disciplinary study. It has also demanded new considerations of complexities and the dense webs of social and material relationships that underlie life on the move. The result of this, however, is that while mobilities research is drawn together by broadly shared concerns and curiosities, after a decade of development it is also marked by significant internal variations. Widespread agreement about basic terms or approaches is not only implausible, but also inappropriate given the emphasis on flexible sets of theoretical and methodological resources. How then might we assess and reflect upon the form this development has taken?

One way of thinking about the development of research over the past decade is to ask whether we can see the emergence of a research field as a result of the efforts of contributors to the mobilities project. By using the term field, we allude to ideas associated with the work of Pierre Bourdieu (1984), as well as a range of other literatures including those on institutional fields (DiMaggio and Powell 1983; Scott 1994). We use the question of the emergence of a field of mobilities research as a way into thinking about two issues. First, we ask the most fundamental question: to what extent might the mobilities project of the past decade be conceptualized as a process of field emergence? In doing this, we consider both the process and politics of emergence and in turn what is at stake in such a project. Second, we consider the potential struggles of the next decade, and draw attention to the important role subfields could play in shaping future interactions and struggles.

Firstly, then, we consider what field emergence might involve. A field is an arena with 'a common meaning system' (Scott 1994, 207–208). As such, a field is a recognized area of social life, with rules, norms and cultures that shape the actions of field members. The strength of a field's meaning system will, however, vary over the course of its emergence. Fligstein and McAdam (2012, 86–87) suggest that emerging fields are occupied by groups that orientate to one another because of shared interests, but which are yet to fully agree on the 'rules of the game'. This explains the relatively weak struggles over what mobilities research is and is not; in the first decade, the 'rules of the game' were in the process of emerging and, as such, no individual or group was able to define what was inside and outside of the field of research. Indeed, studies of attempts to create new fields (Hargadon and Douglas 2001; Navis and Glynn 2010) have shown that the main efforts involve distinguishing the new field from those already existing, whilst also explaining the nature of the work of a new field in a language comprehendible to important audiences. This accounts for the simultaneous efforts to carve out mobilities as a distinctive research arena that focuses on understudied phenomenon, whilst also highlighting the contributions of the field in relation to wider social scientific debates. Justifying and emphasizing the importance of mobilities research as a field has, then, perhaps been the key struggle of the first decade, as a part of efforts to ensure recognition and comprehension within the social sciences.

The primacy of this struggle helps to explain why it is less clear whether mobilities research fulfils Scott's second criterion for the existence of a field. In addition to having 'a common meaning system', Scott suggests that a field's 'participants interact more frequently and fatefully with one another than with actors outside of the field' (1994, 207–208). Due in part to the interdisciplinary nature of mobilities research, this intensity of interaction would be both difficult to assess and potentially inappropriate to expect. Interdisciplinary exchange, after all, benefits from the incomplete overlapping of multiple academic communities, whose differences and debates inform new insights. The increasing acceptance of interdisciplinary work also means that it is now more common for researchers to be intermittently engaged within a range of different fields – some out of interest, and others out of obligation (as for instance in the case of research assessment exercises in the UK or departmental re-organizations where new fields are dictated by others). One might thus challenge the appropriateness of Scott's criterion within an interdisciplinary context, and in a context in which careers are connected to university structures still often centred upon metadisciplines (economics, geography, sociology, etc.) and their fields. That is, when establishing a new field, interactions with actors outside it are incredibly fateful, and have significant consequences for recruiting new participants and establishing the field's importance. Hence, as Fligstein and McAdam (2012) propose, we need to be sensitive to the relationships and interactions that span between fields given the networked nature of contemporary social life.

Nonetheless, even if the potential limits of Scott's definition are accepted, there are suggestions that interactions between mobilities researchers, including through papers, projects and this journal, are becoming increasingly important to agenda setting, theory development and careers within a distinctive area (i.e. field) of academic life. Repeated discursive articulations of the 'mobilities paradigm' or 'mobilities turn' within publications, conferences and meetings help to further build up a set of shared understandings of social life and approaches to research. It seems, then, that a field of mobilities research is repeatedly discussed (Adey et al. 2014b; Cresswell 2011, 2012, 2014; Merriman 2015) and is emerging as a fateful 'space' for its participants. What then are the implications of such a trajectory of development for the next 10 years?

Maturation and New Challenges

The literature on fields points to how, after emergence and as maturation occurs, new considerations are likely to come into play. Fligstein and McAdam (2012, 88) show that in the maturation period roles are carved out for different actors and more sustained attempts are made to create a framework that allows those inside the field to understand the 'rules of the game'. As a part of this coalescence, power relations begin to develop. Those in 'incumbent' positions who play a powerful role in defining the field sit alongside 'challengers' who may have different views of what the 'rules of the game' should be, but perhaps not the resource or motivation to question the emergent consensus. Hence, moving into the second decade of an explicitly acknowledged mobilities turn, one would expect struggles between participants and non-participants of the field (i.e. over what mobilities research is and why it is distinctive) to fade in importance relative to the struggles and competition between mobilities researchers. This could, for example, lead to the emergence of some 'central features' (Urry 2007, 46) and 'lineaments' (Urry 2007, 59) of the field. In any case, the potential for struggle raises important questions about the nature of competition and power within fields, which seem very relevant for the next decade.

Bourdieu's (1984) discussion of fields, and in particular his use of games as a metaphor for the struggles and competition that go on within them, provides a useful starting point for examining such questions. For Bourdieu, interactions within fields are oriented towards increasing one's relative power and prestige compared to other members. Fields thus involve competitions for what MacIntyre calls 'external goods' such as power, prestige or money, which once obtained 'are always some individual's property and possession' (1985, 190). The competition for these goods, what Bourdieu would understand in terms of capitals, is oriented towards individual gains, but based upon a shared understanding of, and commitment to, the values and stakes they reinforce.

Yet, as MacIntyre highlights, external goods are not the only ones to which power and politics can be oriented. In the course of striving for shared standards and purposes, people also realize 'internal goods' that are 'the outcome of competition to excel, but … [whose] achievement is a good for the whole community who participate in the practice' (MacIntyre 1985, 190–191). Competition thus also begets collective gains, such as increases to shared knowledge or the development of shared methods. Whereas the accrual of external goods can be easily understood and compared across different fields, only experts within a practice can judge the accrual of internal goods, as they understand the practice's unique values and rewards (MacIntyre 1985, 188–189).

How do external and internal goods operate within mobilities research? On the one hand, contributing to mobilities research and its establishment as a field has undoubtedly involved the accumulation of external goods for the academics involved – citations, speaking engagements, prestige and promotion. Yet, the dynamics of these goods will change as the field matures, since it can be easier to gain prestige from establishing a new field than from excelling within an existing one. One would expect these goods to become more varied and competitive within the mobilities community, in such a way that positioning within it and not only within broader social scientific fields is clearly at stake. On the other hand, however, there are internal goods that arise from interactions within the mobilities community – encountering new ideas, having a diverse interdisciplinary forum to explore, engaging with new networks of people. These internal goods are not entirely separable from the external ones, which inevitably relate to internal excellence (Knight 1998, 113), but serve to make competition and community interaction inherently rewarding.

The comparative privileging of internal and external goods is important to reflect upon. As Knight highlights, MacIntyre's discussion of goods takes on important political implications when he connects it to the institutions that organize and help sustain practices: 'these indispensable institutions constantly threaten to corrupt practices and demoralize practitioners, to subordinate pursuit of internal goods to that of external goods' (Knight 1998, 11). For MacIntyre, this is extremely problematic. His emphasis upon internal goods of excellence is thus part of a broader move to foreground them within practices, institutions and politics. Focusing too closely upon fields and their struggles for external goods can be politically problematic because it suggests that the rewards of competition 'naturally' accrue to individuals, and positioning is thus an individual responsibility and struggle. In this context, Warde's (2004, 25) suggestion that discussions of fields be supplemented by considerations of practices, which are 'cooperative, pluralistic and orientated towards internal goods', is helpful. It provides space to discuss the relationship between internal and external goods, as well as political struggles over the extent to which practices contribute to individual or collective gains.

This discussion suggests the importance of ongoing academic practices in shaping what the next decade of mobilities research will become. We argue that how field interactions are structured will be crucial to the flourishing of mobilities research in the future, particularly as themes and methods of mobilities research become even more established and perhaps less strongly linked to external goods. Critical reflection on and interventions into the structures and practices of the field could ensure that in the next decade it is not only a space to examine topics, theories and methods not legitimated by other disciplines, but that it is also a space in which to cultivate internal goods unique from those in other interdisciplinary fields. In this way, reflecting upon understandings of fields encourages the consideration of not only what mobilities research studies, but how mobilities researchers are engaged in building up shared resources and encouraging collective gains.

One example of how further space for cooperative and creative internal rewards might be fostered relates to the development of sub-fields. One of Fligstein and McAdam's (2012) key arguments is that the nesting of fields is a fundamental feature of social life, and Abbott has made a similar point about the fractal nature of disciplines (2001). For our purpsoses here, we can translate such ideas into thinking about how the mobilities field is already nested within the wider social sciences, but will likely itself splinter into multiple nested sub-fields. What does this imply? A sub-field will, to some extent, share 'rules of the game' with the wider field. So, for instance, we might expect the emphasis on studying mobilities in and of themselves to be shared across sub-fields. But sub-fields will also be differentiated in terms of agendas, power structures and rules. One can imagine how sub-fields could emerge around work that focuses on particular theorists, as a recent special issue on Foucault (Manderscheid, Schwanen, and Tyfield 2014) and papers in this issue (Adey, Sheller) also imply. Since such sub-fields have yet to fully emerge and there is a wide range of possible foci – empirical, theoretical and methodological – this remains an important moment in which to consider the implications of different configurations.

For example, the formation of sub-fields around theoretical traditions could emphasize or reproduce theoretical debates already well rehearsed in other disciplines. Alternately, the formation of sub-fields around empirical topics such as migration or tourism could discourage creative engagement across the boundaries of these areas, which Hui and Johnson and Martin (this issue) have highlighted. One provocation premised on the value of openness would be to pursue less traditional connections – with sub-fields emerging for instance around the 'qualities', 'materialities' or 'events' that Adey et al. draw attention to (2014a). Another would be to form sub-fields around broad problems such as the politics and ethics of diverse mobilities (Cresswell 2010), theorizing mobilities outside the West (Lin, this issue; Sheller, this issue; Adey, this issue), situating embodied mobilities (Laurier et al., this issue, Licoppe, this issue), and systems of heterogeneous mobilities (Reese, this issue; Merriman, this issue). Since overly rigid or isolated sub-fields seem to us likely to stifle the flexibility and creativity that has made mobilities an appealing interdisciplinary field, there is reason to actively cultivate practices and sub-fields that embed collegiality, open-mindedness, and the destabilization of hierarchies between early career and senior academics as fundamental values.

More fundamentally, this relates to the tricky challenge of managing the fine line between coherence and closure. Resisting the urge to have a specific line on what mobilities research is and is not (closure) would seem important if we are to build upon the successes of the past decade. Resisting the entrenchment of a 'canon' or other markers of internal prestige and hierarchy (closure) would also seem important for maintaining the openness and collegiality of the past decade. As the discussion above makes clear, mobilities research is not united by a coherent ontological position (as with for example actor–network theory), or by an established methodological approach (as with anthropology's foregrounding of ethnography), and the creation of a 'canon' could move towards closing down these areas of diverse discussion. Yet, ensuring that a number of core agendas emerge so that diversity does not become overwhelming, or prevent further dialogue and even creative syntheses, also matters. Refinement and specialization can be important for sparking further enrichment and collective rewards. The reflexivity with which we approach questions of coherence or closure, openness or consensus will shape future trajectories, questions and perhaps the 'rules of the game' that unite sub-fields, as they emerge. We therefore argue that in the next decade mobilities researchers will need to attend not only to questions of how mobile worlds can be studied – in which contexts and using which theories and methods – but also to questions of how the field itself is represented, envisioned and ever on the move.

Conclusions

By highlighting questions about the emergence of mobilities as a research field, we have pointed in this introductory paper to the possibility of tensions and conflict in the next decade. The papers in the special issue all in different ways point towards this dilemma. The topics addressed in each could be the basis for the emergence of a mobilities sub-field (around for instance theories of governance, automobility, aeromobility, method, inequality, etc.), with the associated power struggles and search for goods. Or, the topics could be the basis in the next decade for the further development of open discussions about what mobilities research can achieve without settling into expected or entrenched sub-fields. We prefer a vision built on the latter scenario; while convergence and coalescence bring advantages for conceptual development, excitement and contribution, we feel this must not be pursued without a grounding in the interdisciplinary inspiration that has enlivened mobilities research in the last decade, and which encourages an openness to unexpected conversations, sites of study, and overlaps between individual and collective concerns and priorities.

Hence, for us, the primary concern in the next decade should not be what mobilities research is, what it should be or what distinguishes it internally as a series of sub-fields or in relation to other fields. Rather it should relate to the continuation of dialogues about mobilities – broadly writ – in an exciting, open, and creative shared culture that both draws from but also contributes to the wider social sciences. Evolutions within the field will be an important part of this creativity, and are likely to lead to a mix of both closing down (some theoretical coalescence to incrementally develop important agendas) and the maintaining of porosity (a continued commitment to taking mobility seriously as an object of study and drawing on the most appropriate tools to do this as new issues arise). It is incumbent on mobilities researchers, through their practices, to curate such a field.

The importance of engaging with politically important questions such as climate change and inequalities highlight why a mix of selective closing down alongside generous openness seems important: we need both to 'stand on the shoulders of giants' in our theoretical efforts to deal with such issues (for example through theoretically focused sub-fields), but also respond to the ever evolving nature of the challenges before us (requiring the embracing of new approaches). This special issue is, therefore, our salute to the benefits of the approach of the past decade, to all of the questions not yet asked, and to the rich conversations that they might spark in an evolving (and inevitably changing) field in the future.

Acknowledgements

The editors of the special issue would like to thank the Reviewers for their generosity in commenting on all of the submissions and helping enhance each and every paper, including this introduction. As always, Pennie Drinkall provided outstanding assistance in the CeMoRE and Mobilities office. We also thank Jen Southern (Lancaster University) for her input and support in the early stages of developing this special issue.

References

Abbott, A. 2001. *Chaos of Disciplines*. Chicago, IL: University of Chicago Press.

Adey, P. 2008. "Airports, Mobility and the Calculative Architecture of Affective Control." *Geoforum* 39: 438–451.

Adey, P., D. Bissel, K. Hannam, P. Merriman, and M. Sheller. 2014a. "Introduction." In *The Routledge Handbook of Mobilities*, edited by P. Adey, D. Bissel, K. Hannam, P. Merriman, and M. Sheller, 1–20. London: Routledge.

Adey, P., D. Bissel, K. Hannam, P. Merriman, and M. Sheller, eds. 2014b. *The Routledge Handbook of Mobilities*. London: Routledge.

Aldred, R., and K. Jungnickel. 2013. "Matter in or out of Place?" *Bicycle Parking Strategies and their Effects on People, Practices and Places, Social & Cultural Geography* 14 (6): 604–624.

Anim-Addo, A., W. Hasty, and K. Peters. 2014. "The Mobilities of Ships and Shipped Mobilities." *Mobilities* 9 (3): 337–349.

Bajc, V., S. Coleman, and J. Eade. 2007. "Introduction: Mobility and Centring in Pilgrimage." *Mobilities* 2 (3): 321–329.

Barker, J., P. Kraftl, J. Horton, and F. Tucker. 2009. "The Road Less Travelled – New Directions in Children's and Young People's Mobility." *Mobilities* 4 (1): 1–10. doi:10.1080/17450100802657939.

Basu, P., and S. Coleman. 2008. "Introduction: Migrant Worlds, Material Cultures." *Mobilities* 3 (3): 313–330.

Benson, M. 2011. "The Movement Beyond (Lifestyle) Migration: Mobile Practices and the Constitution of a Better Way of Life." *Mobilities* 6 (2): 221–235.

Bissell, D. 2009. "Conceptualising Differently-Mobile Passengers: Geographies of Everyday Encumbrance in the Railway Station." *Social & Cultural Geography* 10 (2): 173–195.

Bissell, D. 2010. "Passenger Mobilities: Affective Atmospheres and the Sociality of Public Transport." *Environment and Planning D: Society and Space* 28: 270–289.
Blok, A. 2010. "Mapping the Super-Whale: Towards a Mobile Ethnography of Situated Globalities." *Mobilities* 5 (4): 507–528.
Böhm, S., C. Jones, and C. Land. 2006. "Part One Conceptualizing Automobility: Introduction: Impossibilities of Automobility." *The Sociological Review* 54 (s1): 1–16.
Bourdieu, P. 1984. *Distinction: A Social Critique of the Judgement of Taste*. Translated by R. Nice. London: Routledge.
Budd, L. C. S. 2011. "On Being Aeromobile: Airline Passengers and the Affective Experiences of Flight." *Journal of Transport Geography* 19: 1010–1016.
Burrell, K. 2008. "Materialising the Border: Spaces of Mobility and Material Culture in Migration from Post-socialist Poland." *Mobilities* 3 (3): 353–373.
Büscher, M., and J. Urry. 2009. "Mobile Methods and the Empirical." *European Journal of Social Theory* 12 (1): 99–116.
Büscher, M., J. Urry, and K. Witchger, eds. 2011. *Mobile Methods*. London: Routledge.
Conley, J., and A. T. McLaren, eds. 2009. *Car Troubles: Critical Studies of Automobility*. Farnham: Ashgate.
Conradson, D., and A. Latham. 2007. "The Affective Possibilities of London: Antipodean Transnationals and the Overseas Experience." *Mobilities* 2 (2): 231–254.
Conradson, D., and D. Mckay. 2007. "Translocal Subjectivities: Mobility, Connection, Emotion." *Mobilities* 2 (2): 167–174.
Cresswell, T. 2006. *On the Move: Mobility in the Modern Western World*. London: Routledge.
Cresswell, T. 2010. "Towards a Politics of Mobility." *Environment and Planning D: Society and Space* 28 (1): 17–31.
Cresswell, T. 2011. "Mobilities I: Catching up." *Progress in Human Geography* 35 (4): 550–558.
Cresswell, T. 2012. "Mobilities II: Still." *Progress in Human Geography* 36 (5): 645–653.
Cresswell, T. 2014. "Mobilities III: Moving on." *Progress in Human Geography* 35 (5): 712–721.
Cresswell, T., and P. Merriman, eds. 2011. *Geographies of Mobilities: Practices, Spaces, Subjects*. Farnham: Ashgate.
Cwerner, S., S. Kesselring, and J. Urry, eds. 2009. *Aeromobilities*. New York: Routledge.
DiMaggio, P. J., and W. W. Powell. 1983. "The Iron Cage Revisited: Institutional Isomorphism and Collective Rationality in Organizational Fields." *American Sociological Review* 48 (2): 147–160.
Farías, I. 2010. "Sightseeing Buses: Cruising, Timing and the Montage of Attractions." *Mobilities* 5 (3): 387–407.
Featherstone, M., N. Thrift, and J. Urry, eds. 2005. *Automobilities*. London: Sage.
Fincham, B., M. McGuinness, and L. Murray, eds. 2010. *Mobile Methodologies*. Basingstoke: Palgrave Macmillan.
Fligstein, N., and D. McAdam. 2012. *A Theory of Fields*. Oxford: Oxford University Press.
Fortier, A. M., and Lewis, G. 2006. "Mobilities." *Editorial: Migrant Horizons* 1 (3): 307–311.
Gill, N., J. Caletrío, and V. Mason. 2011. "Introduction: Mobilities and Forced Migration." *Mobilities* 6 (3): 301–316.
Hannam, K., M. Sheller, and J. Urry. 2006. "Editorial: Mobilities, Immobilities and Moorings." *Mobilities* 1 (1): 1–22.
Hargadon, A. B., and Y. Douglas. 2001. "When Innovations Meet Institutions: Edison and the Design of the Electric Light." *Administrative Science Quarterly* 46 (3): 476–501.
Hein, J. R., J. Evans, and P. Jones. 2008. "Mobile Methodologies: Theory, Technology and Practice." *Geography Compass* 2 (5): 1266–1285.
Hui, A. 2012. "Things in Motion, Things in Practices: How Mobile Practice Networks Facilitate the Travel and Use of Leisure Objects." *Journal of Consumer Culture* 12 (2): 195–215.
Hui, A. 2013. "Moving with Practices: The Discontinuous, Rhythmic and Material Mobilities of Leisure." *Social & Cultural Geography* 14 (8): 888–908.
Jensen, O. B. 2006. "'Facework', Flow and the City: Simmel." *Goffman, and Mobility in the Contemporary City, Mobilities* 1 (2): 143–165.
Jensen, O. B. 2009. "Flows of Meaning, Cultures of Movements – Urban Mobility as Meaningful Everyday Life Practice." *Mobilities* 4 (1): 139–158.
Jensen, O. B. 2010. "Negotiation in Motion: Unpacking a Geography of Mobility." *Space and Culture* 13 (4): 389–402.

Jensen, O. B., M. Sheller, and S. Wind. 2014. "Together and Apart: Affective Ambiences and Negotiation in Families' Everyday Life and Mobility." *Mobilities*. Accessed March 18, 2015. http://dx.doi.org/10.1080/17450101.2013.868158

Kaufmann, V., M. M. Bergman, and D. Joye. 2004. "Motility: Mobility as capital." *International Journal of Urban and Regional Research* 28 (4): 745–756.

Kidder, J. L. 2009. "Mobility as Strategy, Mobility as Tactic: Post-Industrialism and Bike Messengers." In *The Cultures of Alternative Mobilities: Routes Less Travelled*, edited by P. Vannini, 177–191. Surrey: Ashgate.

Knight, K., ed. 1998. *The MacIntyre Reader*. Notre Dame: University of Notre Dame Press.

Larsen, J. 2008a. "Practices and Flows of Digital Photography: An Ethnographic Framework." *Mobilities* 3 (1): 141–160.

Larsen, J. 2008b. "De-Exoticizing Tourist Travel: Everyday Life and Sociality on the Move." *Leisure Studies* 27 (1): 21–34.

Larsen, J., J. Urry, and K. Axhausen. 2006. *Mobilities, Networks, Geographies*. Aldershot: Ashgate.

Lassen, C. 2006. "Aeromobility and Work." *Environment and Planning A* 38: 301–312.

Licoppe, C. 2009. "Recognizing Mutual 'Proximity' at a Distance: Weaving together Mobility, Sociality and Technology." *Journal of Pragmatics* 41: 1924–1937.

MacIntyre, A. 1985. *After Virtue: A Study in Moral Theory*. 2nd ed. London: Duckworth.

Manderscheid, K., T. Schwanen, and D. Tyfield. 2014. "Introduction to Special Issue on 'Mobilities and Foucault." *Mobilities* 9 (4): 479–492.

McHugh, K. E. 2009. "Movement, Memory, Landscape: An Excursion in Non-representational Thought." *GeoJournal* 74: 209–218.

Merriman, P. 2015. "Mobilities I Departures." *Progress in Human Geography* 39 (1): 87–95.

Navis, C., and M. A. Glynn. 2010. "How New Market Categories Emerge: Temporal Dynamics of Legitimacy, Identity, and Entrepreneurship in Satellite Radio, 1990–2005." *Administrative Science Quarterly* 55 (3): 439–471.

Nowicka, M. 2007. "Mobile Locations: Construction of Home in a Group of Mobile Transnational Professionals." *Global Networks* 7 (1): 69–86.

Ren, C. 2011. "Non-human Agency, Radical Ontology and Tourism Realities." *Annals of Tourism Research* 38 (3): 858–881.

Richardson, T. 2013. "Borders and Mobilities: Introduction to the Special Issue." *Mobilities* 8 (1): 1–6.

Scott, W. R. 1994. "Conceptualizing Organizational Fields: Linking Organizations and Societal Systems." In *Systems Rationality and Parcial Interests*, edited by J. Derlien, U. Gerhardt, and F. Scharpf, 203–221. Baden: Nomos.

Sheller, M. 2014. "Sociology after the Mobilities Turn." In *The Routledge Handbook of Mobilities*, edited by P. Adey, D. Bissel, K. Hannam, P. Merriman, and M. Sheller, 45–54. London: Routledge.

Sheller, M., and J. Urry. 2006. "The New Mobilities Paradigm." *Environment and Planning A* 38: 207–226.

Spinney, J. 2011. "A Chance to Catch a Breath: Using Mobile Video Ethnography in Cycling Research." *Mobilities* 6 (2): 161–182.

Urry, J. 2000. *Sociology beyond Societies: Mobilities for the Twenty-first Century*. London: Routledge.

Urry, J. 2007. *Mobilities*. Cambridge: Polity Press.

Vannini, P. 2011. "The Techne of Making a Ferry: A Non-representational Approach to Passengers' Gathering Taskscapes." *Journal of Transport Geography* 19: 1031–1036.

Warde, A. 2004. "Practice and Field: Revising Bourdieusian Concepts." CRIC Discussion Paper. Centre for Research on Innovation & Competition. Accessed March 18, 2015. www.cric.ac.uk/cric/Pdfs/DP65.pdf

Yeoh, B. S. A., and S. Huang. 2010. "Transnational Domestic Workers and the Negotiation of Mobility and Work Practices in Singapore's Home-spaces." *Mobilities* 5 (2): 219–236.

Uneven Mobility Futures: A Foucauldian Approach

MIMI SHELLER

Department of Sociology, Center for Mobility Research and Policy, College of Arts and Sciences, Drexel University, Philadelphia, PA, USA

ABSTRACT *This assessment of past and future directions in mobility research calls for a Foucauldian approach to better understand the apparatus of uneven mobility illustrated via three examples: tourism mobilities and racialized space, geo-ecologies of elite secession, and disease mobilities and quarantine. Building upon an 'archaeological' and 'geneaological' study of territory, communication, and speed, this essay argues for both a deeper historicizing of mobility research in terms of colonial histories, political ecologies, and biopolitics, as well as a deeper excavation of the material resource bases of mobility in extractive industries, military power, and biomobilities of racial formation. Sovereign control over mobility, individual 'disciplined mobility' and counter-mobilities, and the surveillance, securitization, and production of knowledge about mobilities each emerge as fundamental elements for the future history of uneven mobilities.*

Introduction

Mobility may be considered a universal human right, yet in practice it exists in relation to class, racial, sexual, gendered, and disabling exclusions from public space, from national citizenship, and from the means of mobility at all scales. These barriers to access and controls over mobility are implemented via formal and informal policing, borders, gates, passes, clothing, rules, and surveillance systems that limit the right to move, filter entry and exit, and selectively apply the protection of the state (Cresswell 2006, 2010; Adey 2010b; Adey et al. 2014). Even for those within the gates, so to speak, fragmented public services, hostile policing, anti-human design (such as spikes, fences, and barriers), and gentrified city centers often push the poor, the queer, the disabled, and the non-white to the margins (Blomley 2011, 2014; Sawchuk 2014).

Indeed, the government of mobilities is fundamental to the making of classed, racial, sexual, able-bodied, and gendered subjects through what 'moves' (and resting places) are allowed or denied (McKittrick 2006; Cresswell, forthcoming).

In this article, I argue that sovereign control over mobility, individual 'disciplined mobility' (Packer 2003), and the surveillance, securitization, and production of knowledge of mobilities each emerge as fundamental elements in the history of uneven mobilities that can be built upon a Foucauldian approach to territory, communication, and speed (Packer 2013; Manderscheid, Schwanen, and Tyfield 2014). Uneven mobility, in this view, refers first to *a sovereign terrain for movement* in which there are spatial designs, physical infrastructures, and symbolic impediments creating divergent pathways, differential access, and control architectures for partial connectivity and bypassing; second, it refers to the means or modes of movement that have a greater or lesser degree of ease, comfort, flexibility, speed, and safety, and thus signals *the disciplining of mobile subjects* through affective experiences of moving with more or less friction, noise, danger, fear, or turbulence; and third, it refers to the institutional apparatus and communication media that produce *knowledge of such relations of mobility and immobility*, speed and slowness, comfort and discomfort, security and risk. Here, I focus mainly on the first two aspects, saving questions of knowledge production and communication for future work.

In line with Packer's view of media studies more generally, future mobilities research in a Foucauldian vein should combine an 'archaeology' of the specific apparatus of mobility in a specific context with a 'genealogy' of how powers of mobility have been both constituted and resisted (Packer 2013). Despite emphasis on the politics of mobility in work by Cresswell, Adey, Sheller and others, and important theoretical work on power and mobility such as Anne Jensen's attention to urban space and ambiences (Jensen 2011) or Joergen Bærenholdt's concept of 'governmobility' (Bærenholdt 2013), mobility researchers have not spent enough time showing how governmobilities orchestrate and choreograph embodied differences of uneven mobility across class, gender, race, ethnicity, nationality, sexuality and physical ability. How have sovereign control and disciplinary systems historically produced differently marked bodies as unequal mobile subjects? How do local, regional, urban, national, and global systems for control over territory, communication and speed produce different bodies and scales of disciplined mobility? And what modes of counter power and 'subversive mobilities' (Shell 2015; Montegary and White 2015) might inform the kinds of moves that can be made to resist, overturn, or escape these governmobilities?

Since the publication of 'The New Mobilities Paradigm' (Sheller and Urry 2006a) and the launch of the journal *Mobilities* with the editorial introduction on 'Mobilities, Immobilities and Moorings' (Hannam, Sheller, and Urry 2006) 10 years ago, my own work has focused on themes of uneven mobility, inequality, and power. However, over this decade, it has also become clearer that we need both a deeper historicizing of mobility research in terms of colonial histories, global geographies, and neoimperialism (cf. Sheller 2003), as well as a deeper ecologizing of the material resource bases of mobility in extractive industries, military power, and biopolitical racialization (cf. Sheller 2014). In the global south, on the margins of global cities, and in the regions that Ferguson (2006) calls the 'global shadows' of the neoliberal world order, the challenges of precarious access to mobility produce the sharpest contours of uneven mobility, although it can be found everywhere.

We do have some starting points to build from. To use the categories that serve to organize the recent *Routledge Handbook of Mobility Research*: the unevenness of mobility may take the form of uneven *qualities* of experience, uneven access to *infrastructure*, uneven *materialities*, uneven *subjects* of mobility, and uneven *events* of stopping, going, passing, pausing, and waiting (Adey et al. 2014). Thus, there are many layers to uneven mobility, and many ways to think about its organization and its systematic spatial, temporal, and personal effects. Yet, many recent contributions to mobility research have turned to a micro-scale of inter-personal interaction and mobile methods that seek to capture fleeting moments of movement, while not being explicit enough about the larger apparatus of power in which these relations of mobility are situated and governed. The turn toward new forms of materialism, more-than-human political ontologies, and relational epistemologies have suggested some promising future directions for mobility research, but also need to be put into conversation with more encompassing Foucauldian historical approaches to the apparatus of mobility.

In this assessment of uneven mobilities and future agendas for mobility research, therefore, I want to focus on three brief examples showing the potential for more Foucauldian histories and genealogies of uneven mobilities. I begin with a discussion of tourism mobilities and racialized space, which highlights the relation between what I call micro-level 'biomobilities' and global geographies of 'play-scapes.' Then I turn to geo-ecologies of elite secession, military power, and logistics, sketching the connection between mobility, territorial control, and uneven resource exploitation. And finally I turn to the problem of disease mobilities and quarantine as seen in the recent Cholera epidemic in Haiti and Ebola outbreak in West Africa. While none of these can be fully elaborated here, in bringing these diverse topics together, I hope this intervention will draw attention to the postcolonial, biopolitical, and geo-ecological aspects of uneven mobilities, which I argue have been under-represented in current mobility research.

Tourism mobilities and racialized space

Mobilities are always supported and enabled by a wide range of socio-technical systems and infrastructures that create particular kinds of mobility spaces, and such spaces reproduce power differentials in the capability for mobility. Tourism is one of the key historical and contemporary systems for producing and enacting differential mobility. Studies of tourist mobilities emphasize that there is an unstable and ever-changing interrelation of places, persons, technologies, and natures connected through performances and practices (Edensor 1998, 2000, 2001; Bærenholdt et al. 2004; Sheller and Urry 2004, 2006b). Those fluid relationships create the complex spatialities and temporalities of contemporary tourism development, one of the key qualities of which is the performance of certain places as 'escapes' from the everyday and from the banality of (sub)urban sprawl.

A Foucauldian approach to tourism mobilities might begin from the technologies that produce the mobile tourist, the tourism service worker, and the tourist destination. Three key apparatus for assembling these mobile subjects and spaces are the all-inclusive resort, the cruise ship, and the private island, each of which requires distinctive modes of mobility to get people and goods there. These spatio-temporal enclaves create special qualities and ambiences of mobility that become part of the tourist experience. Entire regions of the world, from the Caribbean to the Pacific to the Indian Ocean, have been enmeshed in systems of tourism development and marketing that revolve around an imaginary of the tropical resort as a leisure space outside of urban modernity in which non-white service workers serve the needs of visitors, producing uneven mobile subjects (Sheller 2003). The production of luxury resorts and enclave tourism also requires the careful staging of black or brown bodies, to perform certain roles but to disappear when not wanted, in a play of visibility and invisibility (Thurlow and Jaworski 2014).

Air travel is one of the key infrastructures for producing the unevenness of such tourism performances. Peter Adey argues in *Aerial Life* (2010a) that air space has a shape and a geography. Moving through the air is an embodied practice, and is performative of space. Adey traces the imperial-aerial gaze back to the colonizing practices of the British Raj and its early practice of biometrics. Adey suggests that the 'emergence of biometrics in contemporary airports, border zones, security spaces and everyday life sees the systematic use of biological and bodily data as a means to identify and manage risky mobile populations, focusing upon not territory but vectors' (Adey 2010a, 88–89). He shows how the forms of enrollment, self-selection into 'trusted traveller' programs, and passenger responsibility to present the data-ready body, all build on colonial practices of mapping, visioning, and surveying (Amoore 2006; Salter 2008; Amoore and Hall 2009). Thus, surveillance enables (neo)colonial governance not only through the mapping of territories and the regulation of mobilities across borders, but also through the biopolitical management of disciplined and racialized bodies enrolled in self-disclosure at the border check-point, borders which themselves are increasingly mobile (Vukov and Sheller 2013).

Aerial power and the view from the air conflate military power, aerial surveillance, and the racializing gaze of the mobile tourist under regimes of secured air travel (Kaplan 2006, 2011), thus producing privileged and excluded subjects through the event of passing through air space with a quality of ease, or getting stuck in its frictions. The global circulation of elites today is governed through privatized corridors of privileged mobility including fast access 'global entry' lanes at airports, various classes of membership and privileges for air travel, gated and exclusive resorts, private jet travel, and the protection of wealth in offshore tax havens (Birtchnell and Caletrio 2014; Urry 2014b). This may involve perpetuating the *biomobilities* of whiteness insofar as frictions of travel become a mechanism for sorting out racialized bodies (as well as transgendered, disabled, or other 'suspect' bodies) through the uneven micro-mobilities of biometric traces, disallowed bodily performances, and racialized and abilized representations of elite versus subaltern mobilities. The project of bordering often becomes an exercise in racial boundary drawing too, controlling the mobilities of racialized bodies into and out of the nation-state, encouraging the migration of some groups while discouraging the cross-border movement of others (Nicholson and Sheller, forthcoming), and allowing for the transgression of racial boundaries in the liminal space of the holiday, the beach, and the tropical island (Nagel 2003; Vicuña Gonzalez 2015).

Mobility research should have far more to say about the ways in which such travel and leisure destinations are being disembedded from national territories and repackaged not only as unique natural enclaves that are connected to global metropolitan transport, media, and data flows, but also as actively *islanded* spaces set apart for carefully managed racial relations of orchestrated security enabled by private ownership, private infrastructure, and control of others' mobility (Sheller 2009a, 2009b). It is also imperative to think about how military power informs the making of safe spaces for tourism, which in turn may become dangerous spaces of terrorist attack, as occurred for example with the Bali, Indonesia, bombings in 2005, the pirate attacks on tourist beaches near Mombasa, Kenya, in 2014, or the attacks on tourists in Tunisia in 2015. On Vieques Island in Puerto Rico, uneven mobilities also inform how a former US military weapons testing range was converted into an 'untouched' nature preserve and a site for exclusive tourist resorts (Sheller 2007). In Vieques, we can see a particularly acute example of the spatial restructuring of land, infrastructure, and media representations in order to produce a new performance of place – one that draws on uneven mobilities to support tourism rather than militarization, yet implicitly still facilitates US military power by normalizing destructive processes of island occupation, targeting, and weapons testing.

Vernadette Vicuña Gonzalez's fascinating book *Securing Paradise: Tourism and Militarism in Hawai'i and the Philippines* (2013) demonstrates the links between militarism and tourism, which are crucial to many Pacific Islands, but also in the US relation to the Caribbean. She shows how particular forms of aeromobility (such as helicopter tours) and road-building (such as the H-3 Interstate) connect histories of military occupation and the tropical island tourist gaze through embodied mobilities and spatial practices. Helicopter tours of Kaua'i, she argues, 'generate mobilities, fields of vision, and structures of feeling that produce landscapes for tourist pleasure that are profoundly interconnected with past and present military violence'; aerial accessibility of the Edenic tropical island depends upon the '*interoperability* of technologies, visual systems, and logics of tourism and militarism upon which helicopter tours rely' (Vicuña Gonzalez 2013, 149). Vicuña Gonzalez suggests that 'The tourist was cast as the innocent subject of leisure whose right to move freely and safely exemplified the ideologies of neoliberal governance' (Vicuña Gonzalez, 2013, 218). Indigenous lands are seized, local mobilities are severely restricted, and access is controlled for the benefit of a few.

The same might be said of urban centers of global cities; here too the practices of conspicuous consumption and unfettered mobility of elites stand in 'obscene contrast' to those evicted to the margins and peripheries, unable to access the city, which has increasingly become a staged spectacle of elite privilege and tourist consumption. In the glistening metropolises of fortified corporate skyscrapers (Kaika 2011; Graham, forthcoming) and commodified live/work/play-scapes (Sheller and Urry 2004; Davis and Monk 2011), there are complex choreographies of embodied mobilities and new forms of architectural stratification. Only disciplined mobile subjects may occupy the best, cleanest, greenest locations and make use of the newest, fastest infrastructures of transport and communication. These uneven terrains of cocooning and vertical mobility bring to the social and political foreground the socio-technical infrastructures pivotal to controlling and governing uneven mobility. They depend not only on the design of the built environment through which (im)mobilities are staged (Jensen 2013), but also on the governmental practices in which delay, exclusion, turbulence, blockage, detention, body search, and disruption are an everyday performance for those who must dwell in, wait patiently, and move uneasily through marginalized spaces seeking livelihoods, passage, and asylum (e.g., Graham 2009; Mountz 2010; Fischer, McCann, and Auyero 2014).

In sum, if we seek to understand the uneven mobilities associated with tourism, we need to trace the historical interplay between various technologies of speed (i.e., from air power to fast-track airport screening), territorial control (i.e., from colonial empires to tourist resorts and private islands), and communication and data sharing (from steamship mail services, to passports, biometrics, and trusted traveler databases). Next, I turn to the problem of sovereign control over elite and subaltern mobilities, and seek to connect processes of 'elite secession' to deeper geo-ecologies of energy use and resource extraction that are dependent on military power.

Geo-ecologies of elite secession and military power

If 'motility' can be defined as 'the manner in which an individual or group appropriates the field of possibilities relative to movement and uses them' (Kaufmann and Montulet 2008, 45), one can begin to conceptualize how elites appropriate not just land and capital, but also possibilities or potentials for movement to places with clean air, water, and food. Historians of technology have shown us how 'infrastructuring' (Star 1999) is an active process of making, embedded in everyday social practices (Guy and Shove 2000), while geographers such as Graham and Thrift (2007) have drawn our attention to the behind-the-scenes yet labor-intensive work of maintenance and repair that gives us the smooth surfaces of 'ready-to-hand' infrastructure such as electricity, communication systems, roadways, and flows of water and waste through pipes. While once developed Western states may have had national visions and urban centralization of such projects, they have become increasingly fragmented and splintered (Graham and Crang 2001).

Here we could think about those who have to wait, in contrast to those who are speeded through. Such systems, Gillian Fuller suggests, rely on the 'everyday infrastucture' of the queue as a kind of 'control architecture' that patterns stop-and-go mobility. There is a daily dance with inherent power relations in the organization of waiting: 'If speed and movement is a commodity then delay is the control,' Fuller argues, such that 'for some, with smart technologies and good salaries, queues can be jumped, for others they are unavoidable.' (Fuller 2014, 212) Having to wait, while others 'speed' past, is a form of power (e.g., experienced daily by car-drivers who jump into the 'fast lane' or get on the 'expressway,' an auto-mobilized metaphor that has been widely taken up to describe elite trajectories of success and advancement).

Waiting implies being stuck, temporally as well as physically (though see the discussion of loitering below). Being stuck also pertains to exposure to disaster risk, toxic pollution, and vulnerability caused by unsustainable and inequitable mobility systems that trap people in fragile environments without food, clean water, sanitation and breathable air, environments from which they cannot escape. Auyero and Swistun (2009), in their account of the Argentine shantytown called Villa Inflamable or 'Flammable,' describe the slow-motion environmental catastrophe of life next to one of the largest petrochemical compounds in Argentina, where a neighborhood is surrounded by oil refineries, unsupervised waste incineration, and landfills, with soil, air, and water contaminated by lead, chromium, benzene, and other chemicals. The opposite of the world of highly mobile elites, this is a place of relegation and waiting, where the poor quality of environment is designed to destroy the biomobilities of living bodies, human and otherwise, even as petrochemicals support the mobile lifestyles of the global elite.

Patterns of climate-related social and environmental collapse are deeply affected by existing global mobility regimes and the uneven distributions of energy, materials, populations, and economic opportunities that they create. As populations find themselves vulnerable to a changing climate and threatened with loss of access to water, energy, or food, issues of security will increasingly come to the fore – including the potential for political unrest and state securitization brought on by climate change and associated urban disasters (Graham 2009). As mobility becomes rationed or far more highly priced due to the politics surrounding the unsustainability of current mobility systems, the inequalities of network capital will be thrown into sharper relief.

Elliott and Urry define network capital as a combination of capacities to be mobile, including appropriate documents, money and qualifications; access to networks at-a-distance; physical capacities for movement; location-free information and contact points; access to communication devices and secure meeting places; access to vehicles and infrastructures; and time and other resources for coordination (Elliott and Urry 2010, 10–11). There is an uneven distribution of these capacities for potential movement in relation to the surrounding physical, social, and political affordances for movement. Here we see how the event of mobility depends not only on the infrastructures that support it (which is so often the focus of more limited approaches to 'sustainable transport'), but also on the materiality of larger assemblages of access and the making of biomobile subjects who can remain connected in a 'disconnected' world (Sheller, forthcoming). Mobility is hence a confluence of geo-ecologies and biopolitics, spanning multiple scales at once.

Displacement following Hurricane Katrina in New Orleans and the humanitarian response in post-earthquake Haiti brought to the fore the astounding interdependence and fragility of the complex mobility systems and infrastructural moorings that make up contemporary global urban geographies (Sheller 2013, forthcoming). Cities suffering catastrophic events are illustrative of how the dynamic intertwining of transportation, communication, provisioning, and scheduling systems can rapidly unravel, and with them the ability of people to stay in place and to live. In the face of mobility systems-failure (and more broadly the impacts of climate change), we see laid bare the institutional scaffolding and regulatory regimes that govern the spatialization of particular territories, including all the gaps and uneven distributions of social rights and network capital that leave some groups most vulnerable to harm (Sheller and Leon, forthcoming).

When we imagine the resilience of transport corridors and mobility systems in the future, we need to consider the 'secession' of elites who are able to provide for their own security and safety through privatized corridors and the accumulation of network capital, while relegating others to situations of slow, encumbered, or vulnerable mobility. 'Secessionist mobility' may be as simple as moving to the automobilized suburbs (Henderson 2006), or as elaborate as owning a private island, using offshore banks, and flying in private jets (Baldacchino 2010; Urry 2014b). Offshoring supports elite secession – including tax havens, free trade zones, export-processing zones, flags of convenience, internet business, shell banks, and so on. Capital is able to move covertly, escaping national territorial legislation and taxation, while labor is fixed in place, with people determined to be 'economic migrants' unable to obtain visas to travel. As John Urry argues, secret offshore worlds and rule-breaking are facilitated by 'various new sociotechnical systems, of container-based cargo shipping; extensive air travel; the internet; huge amounts of car and lorry traffic; new electronic money transfer systems; the growth of taxation, legal, and financial expertise oriented to avoiding national regulations; and the proliferation of many "mobile lives" involving frequent movement across borders' (Urry 2014a, 227).

Access to energy, and the minerals and metals that make up transportation systems, are a crucial dimension of such secessionist mobilities. The potential for mobility is grounded in where energy for transportation is sourced, where it is exported, and who uses most of it. When carbon-hungry kinetic elites resort to the private jet, the helicopter, the high speed train, the yacht, the cocooned limousine, or the semi-militarized sport utility vehicle, they externalize the environmental impacts of high-carbon lifestyles onto others. These exclusive means of mobility also determine spatial development that supports gated enclaves, highrise towers, sanitized pseudo-urban tourist zones, all inclusive resorts with private beaches, spectacles of consumer capitalism, and eco-resorts that are off limits to locals (Sheller 2009a, 2009b; Birtchnell and Caletrio 2014). Kinetic elites can monopolize control over energy, water, and mineral rights, using their power to control global resources that are becoming increasingly scarce.

Unsustainable mobilities also rest crucially on colonial, neocolonial, and neoliberal forms of power that embed racial injustice into these global mobility systems and uneven energy infrastructures. Control of such mobility resources also ensures military supremacy because bombers, drones, armored vehicles, weapons, and satellite guidance systems all need primary metals like aluminum and nickel, while communication systems need silicon and rare earth metals like cobalt and tungsten. Materials and metals are assembled into weapons and vehicles, which in turn need the global flow of oil and gas through pipelines and tankers and logistics hubs to fuel them. Uneven mobilities are therefore geo-ecological at their root, as well as geo-political (Sheller 2014).

More broadly speaking, 'offshoring processes' include the engineering of extra-national quasi-autonomous territories: detention camps like Guantanamo Bay, military bases, world heritage sites, and otherwise autonomous jurisdictions that are lifted out of normal sovereignty (Baldacchino 2010). Such spaces often require special infrastructure to connect them. The current expansion of the Panama Canal, the building of deep-water ports to accommodate the post-Panamax freighters, and the possible building of new logistics hubs in places like the protected area in Jamaica known as Goat Island, site of a planned Chinese logistics hub, all point toward the ways in which flows of global commerce trump local economies and ecosystems, turning nature itself into infrastructure (Carse 2012). Caribbean and Pacific islands are prime locations of US military bases, weapons testing ranges, and oil drilling installations, all of which are crucial military assets that the United States uses to secure its dominance, but are increasingly being encroached upon, especially by Chinese investments in new infrastructure (roads, ports, railways). Such military geographies also produce (and rely on) uneven mobilities, which are closely connected to tourism, as already noted. We need to consider the processes of eviction and expropriation by which once-occupied spaces are unpeopled, 'developed,' abandoned, and then declared 'conservation zones' and represented as 'natural,' 'pristine,' and 'untouched' – ripe for tourism and connected to new chains of logistics.

Deborah Cowen's recent work on logistics rationalities shows that logistics has undergone profound changes since World War II, becoming a driving force of military strategy and tactics, as well as in corporate practices for the global organization of trade (Cowen 2010, 2014b). Logistics, or ('the management of the movement of *stuff*'), is a core system of mobility that supports these uneven geoecologies. Understanding logistics concerns both the infrastructure of global movement of goods and people and the imaginaries and symbolic meanings that drive such organizations of space and time. With the rise of the shipping container, the computer, and satellite communications to manage logistics, new forms of systems thinking have come to the fore, re-shaping labor relations, landscapes, and forms of security. 'Logistics is haunted by its military, imperial and more recent corporate past,' writes Cowen (2014a, 194), entangling supply chain security with border regimes and a recalibration of security. Cowen argues that militarization drives secured enclaves in which military logics re-shape space and employ migrant racialized labor forces without rights of citizenship or collective organization.

Wars in the future will revolve around such appropriation of fundamental resources and control over the logistics systems that allow us to move materials and materiel around the world. Yet, it is exactly such wars and their new forms of weaponry that are depleting the earth's resources and contributing to ecological instability, pollution, and existential threats to human and non-human life, the costs of which are falling most heavily on the poorest parts of the world. A transition toward sustainability of our earth's so-called 'ecosystem services' therefore requires more than changing how much energy we use in everyday life; it more fundamentally requires greater equity in the distribution of network capital and capabilities for mobility, perhaps achievable only through an overthrow of the existing apparatus of mobility that would redistribute – and add procedural and deliberative justice to decisions about – the qualities, events, subjects, materialities, and infrastructures for movement and dwelling. The corollary to this is that a transition toward mobility justice and less uneven distributions of access will require the emergence of a counter-power of capillary mobilities in order to ensure the ecological sustainability and resilience of planetary systems for life. In the next section, I turn to a final uneven mobility, the governance of disease.

Disease mobilities and the geography of containment

If mobile materialities, both human and non-human, inform the making of uneven spatialities and differential mobile subjects, what happens when virulent disease erupts onto the contested stage of sovereign power over territory, communication, and speed? Exposures to disease and variability in human population resistance have long shaped patterns of global mobility, and disease mobility itself has shaped sovereign power and the extension of territorial control. It is well known that the introduction of new diseases assisted in the European conquest of the Americas, and likewise that European vulnerability to tropical diseases long kept white people out of parts of Africa and turned many colonies into so-called 'graveyards.' Disease assisted some of the most infamous military defeats in world history, such as Napoleon's loss of the colony of Saint Domingue (following devastating loss of troops to fever), leading to the independent Republic of Haiti and the Louisiana Purchase by the United States. Disease mobilities have been understood as a significant dimension of more-than-human mobility since the start of the new mobilities paradigm when attention was paid to Foot and Mouth disease, SARS, and the relation between microbial mobilities and jet planes (Sheller and Urry 2006a).

Recent work has begun to connect the study of disease mobilities and transportation to deeper histories of colonial mobility and the contemporary racialization of borders and knowledge (Mavhunga 2014). Mavhunga (2014, forthcoming) shows in his study of human and non-human mobilities in Zimbabwe that disease, mobility, and colonial racism are deeply interconnected. The African traditional knowledge of how to live with Tsetse fly was disrupted by European 'settlers', exposing people and livestock to trypanosomiasis. Europeans appropriated African practical knowledge of how to avoid the disease through the management of mobility, even as they pushed African populations into areas rife with Tsetse. Mavhunga suggests that livestock became 'organic vehicles' for the movement of disease, and that animal and non-human mobilities have been crucial shapers of human mobility, and thus of processes of colonization and resistance (Mavhunga, forthcoming). Furthermore, as Stephanie Lavau argues, viruses like H5N1 avian influenza are both 'travel companions' who hitch a ride on living vehicles such as migratory birds, but they are also a kind of 'companion species' which is ontologically mutable, which is to say it can change form, jump species, and quickly acquire genetic mutations and antigenic resistance (Lavau 2014, 299). In this sense, 'going viral' is a good metaphor for an uncontainable mobility that has both a capacity to move and a capacity to change.

Yet the question of disease mobilities has not become as prominent in mobility research as one might have expected. In this final section, I therefore want to touch briefly on the recent outbreaks of Cholera in Haiti and Ebola in West Africa, as they are in many ways symbolic of the uneven mobility differentials and viral instabilities of our times. Drawing attention to materiality, environmental affordances, and human-animal-technological assemblages in mobility research, we can interrogate how the uneven mobilities of disease outbreaks assist in the re-making of space and biopolities (i.e., a body politic defined around the biopolitics of who deserves protection, who has immunity, and who should be subjected to quarantine). The geographies of the spread of HIV infection, for example, have been racially inflected in terms of their representation in the media (which famously initially blamed Haitians for the introduction of HIV into the United States) and in terms of the distribution of human vulnerabilities confounded by racial, class, and sexual disparities. So too do the vectors of Cholera and Ebola take advantage of the uneven affordances enabled by racial, national, and class (infra)structurings of space. Human responses are also shaped by representations of race and space, whether Western fears of 'darkest Africa,' or African villager's fear of foreign health workers. Infectious diseases get a grip on humanity through the frictions of uneven mobility, exploiting some of us more than others as vehicles for going viral.

An easily transmissible (i.e., virulent, to use a viral vocabulary) strain of Cholera was introduced into Haiti about nine months after the 2010 earthquake, with all evidence pointing toward it being introduced by UN peace-keeping forces from Nepal. They traveled from an active Cholera-epidemic straight to a remote rural base with a faulty sanitation system spilling directly into the Artibonite watershed in central Haiti's main agricultural region (Katz 2014). To date more than 8500 people have died and more than 700,000 have been infected in Haiti, with cases also spreading to the Dominican Republic, and a few carried by air travel beyond the island (PAHO 2014). The UN has refused to accept responsibility, despite extensive evidence and several lawsuits, and only recently announced that it would help address the extremely weak water and sanitation systems in Haiti which has contributed to the spread of the disease. Yet, it is in part the very global mobilization of humanitarian missions that exposes vulnerable populations to disease outbreaks, just as colonizing expeditions once brought disease into unexposed indigenous populations (Sheller 2013, forthcoming).

The mobilities of world governing bodies and humanitarians to 'do good' often mask the implicit military occupations and their associated uneven mobilities that affect the occupied populations. Various kinds of breakdowns in governance often produce new patterns of mobilities: of soldiers, refugees, humanitarian assistance, engineers, researchers, money, materials, but also of viruses, bacteria, and other pathogens which accompany these vehicles. The mobility of UN troops and various non-governmental organizations, for example, has also been linked to increases in prostitution and sex trafficking in areas of the world that they are allegedly aiding, and this in turn is linked to the spread of HIV. Access to air travel, airports, and military air transport underpin these global movements, along with smooth highways and deep sea ports. Peacekeepers protect themselves behind razor-wired encampments and use armored vehicles to move through local neighborhoods; when trouble arises, they also can fall back on diplomatic immunity or simply leaving a country to avoid prosecution for sex crimes, or in this case to avoid taking responsibility for introducing Cholera in a country where people enjoy neither viral immunity nor diplomatic immunity, suffering a lack of biopolitical munitions.

As newly virulent diseases mutate, jump species, and circulate in new ways assisted by global travel, the securing of airports and air travel have become crucial concerns in the containment of disease, and infrastructures for the protection of the qualities of safety enjoyed by the mobile elite. As the world faces new kinds of disease outbreaks and potential epidemics such as avian flu, drug-resistant Tuberculosis, multi-resistant bacteria, and most recently the terrible outbreak of Ebola in West Africa, we have seen the emergence of new global regimes for managing mobilities – especially as control of borders, airports, and entire urban zones become key points of containment. As Ebola spread from Guinea, to Sierra Leone and Liberia, it became increasingly evident that procedures to stop the spread of infection were not working. Fear of the disease rapidly led to the closure of borders, the removal of white bodies from the affected area, and a kind of cordon sanitaire around populations treated as dangerous and primitive. Africans and others around the world were appalled when the first two white Americans infected with the Ebola virus were flown out of the country and given a secret antibody serum at a hospital in Georgia, while African victims were being left to die in isolation. Even though some countries canceled flights to African countries that had not even been affected (e.g., Korean Air stopped flights to Kenya where the disease had not reached), and introduced new screening processes to detect sick passengers, air passengers nevertheless transmitted the virus into Lagos, Nigeria, and Dakar, Senegal.

When human populations become vectors of new microbial and viral threats, the medieval practice of the 'cordon sanitaire' is reintroduced. This became particularly controversial in Liberia, where the disease ran out of control in the capital Monrovia, overwhelming health facilities, while the attempt at draconian quarantine of entire urban neighborhoods met with resistance. How will quarantine be used in the future to immobilize threats to global health? And how will this reflect racialized processes of the sorting of populations into the safe and the dangerous, and identification of racialized territories as unsafe and disease ridden? Few commentaries on the Ebola outbreak have noted the economic and geo-political context that exists in Guinea, Liberia, and Sierra Leone, and how this has contributed to the weak health infrastructure and lack of rural and urban poor access to health care. Is it any coincidence that viral attack on the human species arises in countries at the bottom of many global indexes of health, wealth, maternal mortality, child life expectancy, and income distribution? Let us briefly bring into focus the uneven mobilities of Guinea, where the recent Ebola outbreak began.

Guinea holds two-thirds of the world's reserves of bauxite, as well as major deposits of iron ore, gold, diamonds, and other metals crucial to global mobility infrastructures. The mining sector provides 95% of the country's export earnings, yet Guinea remains one of the poorest countries in Africa, in many ways exemplifying the problems of transnational extractive industries (EITI 2014, Sheller 2014). In the 1990s, under International Monetary Fund and World Bank structural adjustment agreements, the government began to reduce the state sector through extensive privatization and a 1998 investment code opened up the country to foreign investment and allowed for repatriation of profits. Transnational mining corporations such as Rio Tinto and Rusal vied to invest there, and allegations of corruption were rife, amidst government instability and several coups. In 2007 and 2009, armed security forces killed hundreds of protestors and raped dozens of women in public. Nevertheless, 1.5 billion euros ($1.93 billion) of contracts were signed by the former military junta along with 2.1 billion in debt relief from the IMF and World Bank. In 2009, the Chinese International Fund negotiated a $7 billion investment in infrastructure projects including ports, railway lines, and power plants, in exchange for becoming a 'strategic partner' in all mining projects, in what was to be the largest private investment in Sub-Saharan Africa ever.

Does this mobility of global capital and lust for metals to build the global economy, with all of its spatial infrastructures for uneven mobilities, have anything to do with the outbreak of Ebola? Does the extreme poverty of people in remote parts of Guinea (and in the slums of Monrovia) reflect the government's lack of investment in the country while it was lured by huge foreign infrastructure investments in ports and roads to serve the mining sector? When humanitarians fly in to assist the people in Guinea, or Sierra Leone and Liberia, are they even aware of the role of transnational capital in the country? And what happens to 'quarantined' containment areas when even Doctors without Borders will not go there, when hygiene and sanitation equipment is not available, and when local medical staff are dying? Michael Osterholm, Director of the Center for Infectious Disease Research and Policy at University of Minnesota published an opinion piece in the *New York Times* noting that the virus could mutate quickly toward air-borne transmission, posing a threat to humanity, and calling for the United Nations Security Council to seize control of the situation by using military flights to establish supply chains of material for infection control as well as food and water for quarantined areas (Osterholm 2014). What goes unstated in his solution is that this is premised on a total quarantine of these countries and UN control of air travel, taking sovereign power over mobility and isolating particular African populations from the rest of humanity.

Even at their best, humanitarian responses involve uneven mobilities. Humanitarian responders to disaster often bring with them a high degree of network capital, including network connectivity (Sheller, forthcoming). Despite the extensive digital connectivity of global health and humanitarian organizations, they have been unable to bring enough physical supplies and personnel to the affected areas. But this lack of on-the-ground supplies was already produced by the logistics chains that ignored local economies while supplying global mining and extraction economies in West Africa. As I have argued in the context of post-earthquake Haiti, the very same digital connections and logistical efforts that enact processes of rescue, rebuilding, and post-disaster recovery simultaneously produce disconnections, demobilizations, and what I have theorized as an 'islanding effect' (Sheller 2013). Ironically, the new aerial surveillance technologies and GIS integration into real-time mapping for disaster management reinforce the same mobility regimes that are used for border control, security management, migrant interdiction, and perhaps in the future increasingly for quarantine control. Through the disaster response itself, I have argued, rapid deployment of these new technological infrastructures supports the deepening of social inequalities, the re-making of uneven spatialities, and the re-creation of subjects with differential network capital. Jonathan Katz likewise argues that the humanitarian response to the Haitian earthquake was like a big truck going by, and it left behind a disaster (Katz 2014), an *a propos* metaphor for uneven mobility. The Cholera epidemic has been one crucial part of that disastrous aftermath of the uneven mobilities of imperialism. The Ebola outbreak and responses to it are an even more extreme case of uneven mobilities, in which differentiated mobile subjects are exposed in different ways to quarantine and to death itself.

Conclusion

In this brief overview of some future agendas in mobility research, I have tried to weave together several issues that stand out for me as key problems of uneven mobility: the qualities, spaces and performances of tourism and elite secession; the production of more and less mobile subjects through a global biopolitics of racial space; the infrastructuring processes that depend on the materialities of resource extraction, militarization, and viral travels. Within a Foucauldian approach, each of these topics also calls for greater 'genealogical' attention to the longer histories of the colonial, imperial, and military apparatus that inform the government of contemporary global mobilities, as well as resistance against them; and 'archaeological' attention to the deeper geo-ecologies of resource extraction and energy use that support the splintered infrastructures and uneven materialities of mobility. The political organization of the mobility of matter and bodies hinges on a combination of micro biomobilities and macro geoecologies that together contribute to producing and naturalizing the unsustainability of energy usage and ecosystem destruction associated with contemporary forms of life. But such technologies of power might also be turned toward other mobility futures.

While a great deal of attention has been given in recent mobility research to questions of urban sustainability and the new mobilities of global elites and global cities in the twenty-first century, we need to do a better job of tracing the interconnected world-wide historical patterns informing today's uneven mobilities, including their military, colonial, and racial histories. I have argued that a Foucauldian approach to the fundamental variables of territory, communication, and speed could shed light on uneven mobilities. Uneven mobilities are crucial to the forms of elite secession, territorial appropriation, and resource extraction that built colonial worlds and were braided into ongoing forms of power that underwrite the mobile production of contemporary global inequalities.

However, uneven mobility is always contested through a micro-politics of proximity and crowding, gazing and hiding, spying and counter-surveillance. On the one hand, we see a global mobile elite whom Anthony Elliott argues have distinct social practices that cultivate 'detached engagement, speed, networked possibilities, distance from locality and the mapping of escape routes' (Elliott 2014, 37). On the other hand, at the opposite extreme we see the emergence of 'informal infrastructures' for survival, which Colin McFarlane and Alex Vasudevan describe as 'a set of *mobile planning practices* generative of another urbanism' which is 'increasingly dwelt through multiple mobilities shaped by uneven relations of power, resource, and knowledge' (2014, 256–257). Through practices such as squatting or building self-managed water systems, people living in informal settlements create 'autonomous geographies' for an 'insurgent urbanism' that is itself 'continuously mobile' (McFarlane and Vasudevan 2014, 261–262). Dewsbury (2014, 430) refers to the loiterer as 'a heterodox figure for our times, a schizo-analytic soft subverter (after Guattari, 2009).' Can the production of loitering, unexpected delays, mixed crowds, interstitial publics, and newly punctuated temporalities leverage uneven mobilities for a politics of empowering the marginalized?

As seen in the Occupy movement, Critical Mass bike rides, and so many other radical social movement traditions, the blockage of routine movement can create a politics of occupation as collective presence in public space that disrupts 'normal' mobility spaces (Furness 2010). There are also 'vital frictions' that take place within the uneven terrain of micro-mobilities, as diverse people pass on a sidewalk (Blomley 2014), as they mix, slow and pause for social encounters of conviviality. Even riding the metro in a city such as Delhi may become a site of negotiating space and subjectivity, where 'affective responses' reflect 'interacting fields of power in the city' and 'disciplining codes' of conduct and civility (Butcher 2011, 238). These contrasts between the smooth mobilities of global elites and the striated mobilities of informal infrastructures of survival urbanism will only grow in importance in the face of climate change and its imminent disruptions of mobility systems.

There is something in common between the limited-access exclusion zones of the military enclave or secret detention center turned into a tourist resort, the urban neighborhoods targeted as blighted and given over to 'urban renewal' via gentrification, and the indigenous lands and islands whose populations were wiped out by imported diseases in the name of development: in each case the violence of depopulation is masked by the arrival of the innocent settlers, whose access to mobility, to resources, to energy, and to their own pathogens and immunities, delivers them to the new territory. Where informal urbanism arises, or squatters, or anti-capitalist movements, or community mesh networks, or highway blockades, we might consider how their deployment of various kinds of friction, loitering, and sticky occupation serve as forms of counter-power against elite mobilities.

Human rights to clean air, water, and shelter are now monetized and privatized as 'ecosystem services' which may or may not be available to entire populations and urban zones. Nature may be militarized and weaponized (Hamblin 2013) as site of utter destruction (e.g., entire forests defoliated by Agent Orange, oil fields set on fire, chemical warfare and biological weapons, causing 'natural' disasters and changing climate itself) and logistics may be turned toward the denial of the infrastructures that support life (e.g., military blockades of ports, refugee boats sunk, and cross-border movement of goods and energy disrupted). These are some of the crucial uneven mobilities of the future, and should be the subject of mobility research as it moves into its next decade. It is incumbent upon us to explore more thoroughly the relation between such unequal mobility systems and uneven spatialities both as forms of biopolitical governance of embodied mobile subjects and as sites of potential resistance and critical leverage.

Disclosure statement

No potential conflict of interest was reported by the author.

References

Adey, P. 2010a. *Aerial Life: Spaces, Mobilities, Affects*. Chichester: Wiley-Blackwell.

Adey, P. 2010b. *Mobility*. London: Routledge.

Adey, P., D. Bissell, K. Hannam, P. Merriman, and M. Sheller, eds. 2014. *The Routledge Handbook of Mobilities*. London: Routledge.

Amoore, L. 2006. "Biometric Borders: Governing Mobilities in the War on Terror." *Political Geography* 25: 336–351.

Amoore, L., and A. Hall. 2009. "Taking Bodies Apart: Digitized Dissection and the Body at the Border." *Environment and Planning D: Society & Space* 27 (3): 444–464.

Auyero, J., and D. A. Swistun. 2009. *Flammable: Environmental Suffering in an Argentine Shantytown*. Oxford: Oxford University Press.

Bærenholdt, J. O. 2013. "Governmobility: The Powers of Mobility." *Mobilities* 8 (1): 20–34.

Bærenholdt, J. O., M. Haldrup, J. Larsen, and J. Urry, eds. 2004. *Performing Tourist Places*. Avebury: Ashgate.

Baldacchino, G. 2010. *Island Enclaves: Offshoring Strategies, Creative Governance, and Subnational Island Jurisdictions*. Montreal: McGill-Queen's University Press.

Birtchnell, T., and J. Caletrio, eds. 2014. *Elite Mobilities*. London: Routledge.

Blomley, N. 2011. *Rights of Passage: Sidewalks and the Regulation of Public Flow*. New York: Routlegde.

Blomley, N. 2014. "Sidewalks." In *The Routledge Handbook of Mobilities*, edited by P. Adey, D. Bissell, K. Hannam, P. Merriman, and M. Sheller, 472–482. New York: Routledge.

Butcher, M. 2011. "Cultures of Commuting: The Mobile Negotiation of Space and Subjectivity on Delhi's Metro." *Mobilities* 6 (2): 237–254.

Carse, A. 2012. "Nature as Infrastructure: Making and Managing the Panama Canal Watershed." *Social Studies of Science* 42 (4): 539–563.

Cowen, D. 2010. "A Geography of Logistics: Market Authority and the Security of Supply Chains." *The Annals for the Association of American Geographers* 100 (3): 1–21.

Cowen, D. 2014a. "Logistics." In *The Routledge Handbook of Mobilities*, edited by P. Adey, D. Bissell, K. Hannam, P. Merriman, and M. Sheller, 187–195. New York: Routledge.

Cowen, D. 2014b. *The Deadly Life of Logistics: Mapping Violence in Global Trade*. Minneapolis: University of Minnesota Press.

Cresswell, T. 2006. *On the Move: Mobility in the Modern Western World*. London: Routledge.

Cresswell, T. 2010. "Towards a Politics of Mobility." *Environment and Planning D: Society and Space* 28 (1): 17–31.

Cresswell, T. Forthcoming. "Black Moves." *Transfers: Interdisciplinary Journal of Mobility Studies* 6 (1). Special Issue on Race and Mobility, edited by J. Nicholson and M. Sheller.

Davis, M., and D. Monk, eds. 2011. *Evil Paradises: Dreamworlds of Neoliberalism*. New York: New Press.

Dewsbury, J. D. 2014. "Loiterer." In *The Routledge Handbook of Mobilities*, edited by P. Adey, D. Bissell, K. Hannam, P. Merriman, and M. Sheller, 429–438. New York: Routledge.

Edensor, T. 1998. *Tourists at the Taj: Performance and Meaning at a Symbolic Site*. London: Routledge.

Edensor, T. 2000. "Staging Tourism: Tourists as Performers." *Annals of Tourism Research* 27: 322–344.

Edensor, T. 2001. "Performing Tourism, Staging Tourism: (Re)producing Tourist Space and Practice." *Tourist Studies* 1: 59–81.

EITI (Extractive Industries Transparency Initiative). 2014. "Country Report: Guinea." Accessed September 11. http://eiti.org/Guinea

Elliott, A. 2014. "Tracking the Mobile Lives of Globals." In *Elite Mobilities*, edited by T. Birtchnell and J. Caletrio, 21–39. London: Routledge.

Elliott, A., and J. Urry. 2010. *Mobile Lives*. London: Routledge.

Ferguson, J. 2006. *Global Shadows: Africa in the Neoliberal World Order*. Durham, NC: Duke University Press.

Fischer, B., B. McCann, and J. Auyero, eds. 2014. *Cities from Scratch: Poverty and Informality in Urban Latin America*. Durham, NC: Duke University Press.

Fuller, G. 2014. "Queue." In *The Routledge Handbook of Mobilities*, edited by P. Adey, D. Bissell, K. Hannam, P. Merriman, and M. Sheller, 205–213. New York: Routledge.

Furness, Z. 2010. *One Less Car: Bicycling and the Politics of Automobility*. Philadelphia, PA: Temple University Press.

Graham, S., ed. 2009. *Disrupted Cities: When Infrastructure Fails*. New York and London: Routledge.
Graham, S. Forthcoming. "Vanity, Verticality, and Violence: Reflections on Contemporary Skyscrapers." *International Journal of Urban and Regional Research*.
Graham, S., and M. Crang. 2001. *Splintering Urbanism: Networked Infrastructures, Technological Mobilities and the Urban Condition*. London: Routledge.
Graham, S., and N. Thrift. 2007. "Out of Order: Understanding Repair and Maintenance." *Theory, Culture & Society* 24 (3): 1–25.
Guattari, F. 2009. *Soft Subversions: Texts and Interviews 1977–1985*. Los Angeles, CA: Semiotexte.
Guy, S., and E. Shove. 2000. *A Sociology of Energy, Buildings and the Environment: Constructing Knowledge, Designing Practice*. London: Routledge.
Hamblin, J. D. 2013. *Arming Mother Nature: The Birth of Catastrophic Environmentalism*. Oxford: Oxford University Press.
Hannam, K., M. Sheller, and J. Urry. 2006. "Editorial: Mobilities, Immobilities and Moorings." *Mobilities* 1 (1): 1–22.
Henderson, J. 2006. "Secessionist Automobility: Racism, Anti-Urbanism, and the Politics of Automobility in Atlanta, Georgia." *International Journal of Urban and Regional Research* 30 (2): 293–307.
Jensen, A. 2011. "Mobility, Space and Power: On the Multiplicities of Seeing Mobility." *Mobilities* 6 (2): 255–271.
Jensen, O. B. 2013. *Staging Mobilities*. Abingdon: Routledge.
Kaika, M. 2011. "Autistic Architecture: The Fall of the Icon and the Rise of the Serial Object of Architecture." *Environment and Planning D: Society and Space* 29 (6): 968–992.
Kaplan, C. 2006. "Mobility and War: The 'Cosmic View' of Air Power'." *Environment and Planning A* 38 (2): 395–407.
Kaplan, C. 2011. "The Balloon Prospect: Aerostatic Observation and the Emergence of Militarized Aeromobility." In *Keynote Address at the Mobilities in Motion Conference*. Philadelphia, PA: Drexel University.
Katz, J. M. 2014. *The Big Truck that Went By: How the World Came to Save Haiti and Left Behind a Disaster*. New York: Palgrave Macmillan.
Kaufmann, V., and B. Montulet. 2008. "Between Social and Spatial Mobilities: The Issue of Social Fluidity." In *Tracing Mobilities: Towards a Cosmopolitan Perspective*, edited by W. Canzler, V. Kaufmann, and S. Kesselring, 37–56. Farnham: Ashgate.
Lavau, S. 2014. "Viruses." In *The Routledge Handbook of Mobilities*, edited by P. Adey, D. Bissell, K. Hannam, P. Merriman, and M. Sheller, 298–305. New York: Routledge.
Manderscheid, K., T. Schwanen, and D. Tyfield. 2014. "Introduction to Special Issue on 'Mobilities and Foucault'." *Mobilities* 9 (4): 479–492.
Mavhunga, C. C. 2014. *Transient Workspaces: Technologies of Everyday Innovation in Zimbabwe*. Cambridge, MA: MIT Press.
Mavhunga, C. C. Forthcoming. *The Tsetse Fly Rides Again: Mobility and the African Foundations of Tsetse Science in Zimbabwe*. Cambridge: MIT Press.
McFarlane, C., and A. Vasudevan. 2014. "Informal Infrastructures." In *The Routledge Handbook of Mobilities*, edited by P. Adey, D. Bissell, K. Hannam, P. Merriman, and M. Sheller, 256–264. New York: Routledge.
McKittrick, K. 2006. *Demonic Grounds: Black Women and the Cartographies of Struggle*. Minneapolis: University of Minnesota Press.
Montegary, L., and M. A. White, eds. 2015. *Mobile Desires: The Politics and Erotics of Mobility Justice*. New York, NY: Palgrave Macmillan.
Mountz, A. 2010. *Seeking Asylum: Human Smuggling and Bureaucracy at the Border*. Minneapolis: University of Minnesota Press.
Nagel, J. 2003. *Race, Ethnicity and Sexuality: Intimate Intersections, Forbidden Frontiers*. Oxford: Oxford University Press.
Nicholson, J., and M. Sheller. Forthcoming. "Special Issue on Race and Mobility." *Transfers: Interdisciplinary Journal of Mobility Studies* 6 (1).
Osterholm, M. 2014. "What We're Afraid to Say about Ebola." *The New York Times*, September 11, 2014, Opinion Page.
Packer, J. 2003. "Disciplining Mobility: Governing and Safety." In *Foucault, Cultural Studies and Governmentality*, edited by J. Bratich, J. Packer, and C. McCarthy, 135–161. Albany: State University of New York Press.

Packer, J. 2013. "The Conditions of Media's Possibility: A Foucauldian Approach to Media History." In *The International Encyclopedia of Media Studies*, Vol. 1, edited by A. Vadivia, 2–34. London: Blackwell.
PAHO (Pan-American Health Organization). 2014. "*Epidemiological Update: Cholera, 27 June 2014*," Accessed September 10. http://www.paho.org/hq/index.php?option=com_docman&task=doc_view&gid=25979+&Itemid=999999&lang=en
Salter, M., ed. 2008. *Politics at the Airport*. Minneapolis: University of Minnesota Press.
Sawchuk, K. 2014. "Impaired." In *The Routledge Handbook of Mobilities*, edited by P. Adey, D. Bissell, K. Hannam, P. Merriman, and M. Sheller, 409–420. New York: Routledge.
Shell, J. 2015. *Transportation and Revolt: Pigeons, Mules, Canals and the Vanishing Geographies of Subversive Mobilities*. Cambridge, MA: MIT Press.
Sheller, M. 2003. *Consuming the Caribbean: From Arawaks to Zombies*. London and New York: Routledge.
Sheller, M. 2007. "Re-Touching the 'Untouched Island': Post-Military Tourism in Vieques, Puerto Rico." *Téoros* 26 (1): 21–28.
Sheller, M. 2009a. "Infrastructures of the Imagined Island: Software, Mobilities, and the Architecture of Caribbean Paradise." *Environment and Planning A* 41: 1386–1403.
Sheller, M. 2009b. "The New Caribbean Complexity: Mobility Systems, Tourism and the Re-scaling of Development." *Singapore Journal of Tropical Geography* 30: 189–203.
Sheller, M. 2013. "The Islanding Effect: Post-disaster Mobility Systems and Humanitarian Logistics in Haiti." *Cultural Geographies* 20 (2): 185–204.
Sheller, M. 2014. *Aluminum Dreams: The Making of Light Modernity*. Cambridge: MIT Press.
Sheller, M. Forthcoming. "Connected Mobility in a Disconnected World: Contested Infrastructure in Post-Disaster Contexts." *Annals of the Association of American Geographers*, Special Issue on Geographies of Mobility, edited by M.-P. Kwan.
Sheller, M., and Leon, Y. Forthcoming. "Uneven Socio-Ecologies of Hispaniola: Asymmetric Climate Adaptation in Haiti and the Dominican Republic." *Geoforum*, Special Issue on Climate Justice & the Caribbean, edited by A. Baptiste and K. Rhiney.
Sheller, M., and J. Urry, eds. 2004. *Tourism Mobilities: Places to Play, Places in Play*. London: Routledge.
Sheller, M., and J. Urry. 2006a. "The New Mobilities Paradigm." *Environment and Planning A* 38 (2): 207–226.
Sheller, M., and J. Urry, eds. 2006b. *Mobile Technologies of the City*. London: Routledge.
Star, S. L. 1999. "The Ethnography of Infrastructure." *American Behavioral Scientist* 43 (3): 377–391.
Thurlow, C., and A. Jaworski. 2014. "Visible-invisible: The Social Semiotics of Labour in Luxury Tourism." In *Elite Mobilities*, edited by T. Birtchnell and J. Caletrio, 176–193. London: Routledge.
Urry, J. 2014a. "The Super-rich and Offshore Worlds." In *Elite Mobilities*, edited by T. Birtchnell and J. Caletrio, 226–240. London: Routledge.
Urry, J. 2014b. *Offshoring*. London: Polity.
Vicuña Gonzalez, V. 2013. *Securing Paradise: Tourism and Militarism in Hawai'i and the Philippines*. Durham, NC: Duke University Press.
Vicuña Gonzalez, V. 2015. "Tourism Mobilities, Indigenous Claims, and the Securitization of the Beach." In *Mobile Desires: The Politics and Erotics of Mobility Justice*, edited by L. Montegary and M. A. White, 28–40. New York, NY: Palgrave Macmillan.
Vukov, T., and M. Sheller. 2013. "Border Work: Surveillant Assemblages, Virtual Fences, and Tactical Counter-Media." *Social Semiotics* 23 (2): 225–241.

Emergency Mobilities

PETER ADEY

Department of Geography, Royal Holloway University of London, Surrey, UK

ABSTRACT *This paper explores the relationship between mobilities and emergencies, two concepts that have shared little of the same space in research and critical debate. Emergency is a relatively taken-for-granted part of the political administration of events, life and the production and governance of mobility. Equally, mobilities and immobilities occur in, because of, or through emergency. Some mobilities could certainly be understood* as *emergency because whether in flight or response, emergencies demand highly intensive forms of movement that radically transform one's life chances and quality of life. The paper suggests that particular sets of mobilities occur and are compelled under certain kinds of conditions and forms of governance wielded under emergency politics, its legislation and practices. The paper works to identify several related characteristics of emergency mobility that have begun to be explored within existing literatures, burgeoning areas of enquiry and more conceptual writings, before concluding with a discussion of the implications of these themes for a more modest and provisional understanding of mobilities, emergencies and their governance.*

Introduction

This paper traces out a relationship between mobilities and emergencies, two concepts that have shared little of the same space in research and critical debate. Yet they are not strangers to one another. Even if there is not the space here to pursue a genealogy of emergency and mobility, they are inescapable pairs. Emergency is a relatively taken-for-granted part of the political administration of events, life and the production and governance of mobility. Equally, mobilities and immobilities occur in, because of, or through emergency. Consider, for example, the mobilities erupting from environmental catastrophe, technological failure, disease or civil war. Some mobilities could certainly be understood *as* emergency because whether in flight or response, emergencies demand highly intensive forms of movement that radically transform one's life chances and quality of life. Sometimes they occur across fluid and moving landscapes, landslips, mudslides, earthquakes, tsunamis or just as unstable political-economic, social and domestic circumstances – inhuman natures where the ground is pulled from under one's feet.

Compare, in current times, the legal-juridical declaration of emergency and 'states of emergency' in the context of the Ebola outbreak in Nigeria, Sierra Leone, Liberia and other countries in West Africa. The emergency has produced several circuits of aid, border-points and patient and population transfers, highly inequitable international medivacs of foreign nationals, as well as enforced immobilities through restrictions on international traffic, population curfews and quarantine orders, as seen in Liberia's capital Monrovia, the security lockdown of the West Point slum (BBC News Online 2014) and the 90-day time-delimited state of emergency (LA Times 2014). Similarly, the immigration crisis currently facing European states has seen countries such as Hungary institute a state of emergency in several of its counties as an attempt to govern the movements of migrants through enhanced police powers and new laws.

Either way, emergency, as an exceptional legal-juridical paradigm of government (Agamben 1998), or a far more normal governmental-institutional form, is entirely bound up in mobilities. And yet the relation between mobilities and more nuanced conceptions of emergency and governance are rarely the focus of mobilities research (although see Sheller 2013 on post-disaster humanitarian aid; and for an exception in this journal see Birtchnell and Büscher 2010; Budd et al. 2010; O'Regan 2010). What connects mobilities and emergencies are practices of governance: forms of governance that seek to manage emergency mobilities in various ways. It is this glue between mobilities and emergencies, and the conceptual and empirical divide that currently lies between the ways they are conceived in contemporary scholarship, that this paper seeks to overcome.

This paper suggests that particular sets of mobilities occur and are compelled under particular kinds of conditions and forms of governance wielded under emergency politics (Honig 2013), its legislation and practices. While a variety of perspectives may attend to the consequences of displacement and resettlement and processes of return, mobilities can help us explore how emergencies are governed, freighted with meaning and significance, and lived and experienced. The social science response to Hurricane Katrina has reflected this kind of approach (Bartling 2006; Cresswell 2008; Graham 2005), marking a change not simply in how we understand the social and political construction of a disaster, but that a kind of mobility like emergency evacuation could be seen as much more than a symptom of events, but *productive of the emergency itself*. This paper seeks to move beyond the singular development of these concepts from existing case studies and disciplinary divides, in order to focus more intently on the processes of emergency mobility governance.

First, we begin a survey of the wider mobilities literature, its attention to emergencies and the wider absence of critical conceptual literatures on emergency and governance within this work. In so doing, the paper offers to begin a more developed conceptual discussion concerning mobilities, emergencies and governance by drawing together so far separate bodies of literature.

Led by this discussion, the paper then begins to demarcate several related characteristics of emergency mobility that have begun to be explored within burgeoning areas of enquiry, as a means to develop a theory of emergency mobility. These are in order: anticipation; coordination; the inhuman; mobile machines; absence; difference; and finally, times. The paper concludes with a discussion of the implications of these facets for a more modest and provisional theorisation of mobilities, emergencies and their governance.

Mobilities, Emergencies, Governing

In their agenda-setting editorial from the first issue of this journal, Kevin Hannam, Mimi Sheller and John Urry wrote:

> From SARS and avian influenza to train crashes, from airport expansion controversies to controlling global warming, from urban congestion charging to networked global terrorism, from emergency management in the onslaught of tsunamis and hurricanes to oil wars in the Middle East, issues of 'mobility' are centre-stage. (Hannam, Sheller, and Urry 2006, 1)

It is interesting that the events of global disasters, and the emergency techniques determined to respond to tsunamis or hurricanes, are given as examples of how mobility issues are 'centre stage'. Emergency mobilities appear as exceptional events that we should take notice of, and yet are normal to the precariousness of modern existence that they demand sustained attention.

Despite this illustrative use, emergencies have been somewhat underexamined and subject to theorisation within mobilities research (although see Cook and Butz 2015 on disaster). I do not mean that scholarship has not given significant attention to particular emergencies. As the inaugural editorial of this journal drew on the faltering networks of emergency response during 9/11, to the 'dysfunctional' evacuation of New Orleans during Hurricane Katrina (see e.g. Graham 2005), the 'complex and tightly interlocking systems of mobility, transportation and communication to sustain contemporary urban life' have been revealed by a wide range of authors. These display the ways in which mobility systems might fail in emergency (Graham 2009), and that their failure may well *be* the emergency. The inequalities emergencies produce in mobility experiences, whether in evacuation or resettlement, are common examples, as such studies have tended to become exemplary ciphers of the fragility of mobility as an accomplishment.

More generally, mobility studies has tended to consider life lived on the move in the midst of emergency or crisis; in the ways social inequalities might be reinforced or recalibrated by emergency (Cresswell 2008); in the improvisation and adaption that puts strains on existing mobility practices; and at the level of embodied experience (Samuels 2012). These tend to see mobility more broadly as an outcome of emergency or governance, intended or otherwise, and have a habit of disconnecting the experience of mobility from the ways in which the emergency is governed and managed (for an exception see Sheller 2013).

Other research areas have worked in more depth to conceive of emergency as the political motif of our times. Here the legal-exceptional suspension of the normal rule of law, a 'state of emergency' (Agamben 1998), is an important but not necessarily singular articulation of emergency we could consider which has helped supply analysis of events from extrajudicial killing, heightened security practices, to the manner in which everyday border practices stop, enable and filter out mobile subjects from leaving and entering sovereign states (Amoore 2006). These are routine forms of the exception. Whilst contemporary border practices have been no doubt framed within an emergency politics especially following 9/11, Salter (2008) has argued that other modalities of exception are performed which do not see the simple abeyance of law, but a different normalisation of the ability to decide who falls under, or not, sovereign protections over territory and population. Emergency shapes our ability to allow, deny or expel mobilities across and within borders.

Wider work, however, is beginning to explore how emergencies are assembled not simply through the exception, but at the level of 'mechanisms, techniques and technologies of power' (Foucault 2003: 241). More than a legal-juridical suspension of the normal running of law, or a certain capability of sovereign power, an emergency is a more general and open space-time or interval (Aradau and van Munster 2012) in which threats to life generate a moment when certain sets of action are prescribed, or possible. Contemporary forms of governance, domestic and international, routinely deploy techniques to govern emergency, which do not rely upon the suspension of constitutional powers, and their separations, but statutory powers and responsibilities which are delegated to a roster of organisations who prepare and plan for what to do should emergency threaten (Anderson and Adey 2012; Collier and Lakoff 2008; Grove 2013; O'Grady 2014).

On the other hand, the attention of this research to mobilities is also relatively partial. At one level, as various critics have argued, emergency has had a tendency to be discussed at a high degree of conceptual abstraction, with a preponderance for emergency as exception (see Neocleous 2006 for this critique). At another, it has been predominantly concerned with specific spatio-political formations, such as the border or the camp (Diken and Laustsen 2005; Edkins 2000; Ek 2006), and the subjects caught up within those formations, interned, imprisoned or abandoned, etc.

What seems lacking, I would argue, is sustained theoretical development in order that we might begin to arrive at more sympathetic understanding of emergency mobility in order to take seriously the range of experiences common to emergency mobility; to think far more critically about the ethics and politics of mobility in emergency; to understand the ways in which specific techniques, modalities and practices of governance are deployed; and to ask what is common across a wide array of contexts and situations.

One solution would be to begin to develop an approach that works through both mobility studies, and other areas of academic work more attuned to the politics of emergency we saw above. In the following section, the paper explores several intertwined facets of mobility, emergency and governance in order to develop a theorisation of emergency mobilities.

Governing Emergency Mobilities

We might first make several important assumptions in a theory of emergency mobility. First, the nature of emergency: emergencies can be applied to numerous contexts and situations, or events, such as in response to environmental disasters, chemical explosions, nuclear meltdowns, fires, flooding, war, terrorism, rioting, enemy-alien threat. These kinds of emergencies – named or designated by a range of designating actors who are usually state officials, local government, emergency services – are notoriously mobile, and difficult to predict, spreading like wildfire, cascading across different societal systems (Little 2006) or lurking unseen as 'rising tide' type emergencies. Emergencies may involve uncontrolled populations trying to flee as first, second or different order effects, as well as widespread disruption. Equally, emergencies constitute blocked mobilities. Some are trapped or stranded, or disorientated as to where to go, and are struggling to be set free (Sheller 2013). Emergency may therefore involve qualifiable interruption or disruption to mobility, or prompt and even force other unplanned and desperate ones. In short, emergencies are named or designated, and produce multiple forms of (im)mobility.

Second, mobility comes to constitute the ways that governance responds to emergency, just as the designation of emergency itself may *designate* a series of potential legislative and procedural practices of response. The emergency governance of mobility seeks to organise a series of activities, practices, technologies and representations that work in concert to respond and plan so as to get things moving again, across various scales, some intensely local, others as Sheller (2013) showed in the context of Haiti, are transnational. This may mean the development of multiple techniques, like the creation of emergency plans, usually sequences of action that guide how responders and populations should act, where they should move to and what decisions they should make while bringing people to safety (Adey and Anderson 2011). Almost by definition, the mobilisation of aid and humanitarian organisations means complex distributions of mobilities of people and things to distant places (Calhoun 2010). In other contexts, ambulance, accident, and fire workers move rapidly to the scene. From the complex logistics of the mobilities of aid and medical response (Fassin and Pandolfi 2010), to the movements of emergency services and the evacuation of the vulnerable; in other words, mobilities are often performed as efforts to govern emergency (Ikeya 2003).

Finally, while a notion of emergency mobility may take research beyond exceptional emergency legislation to the techniques and practices used to govern normalised emergency mobilities, it does not mean we should remain closed to the possibility of exception. The governance and organisation of emergency mobilities through the normal deployment of the mechanisms, techniques and technologies of power, may allow, in some circumstances, for the temporary return to the imposition of a more Agambenian 'state of exception' in some moments.

Thus, a conceptualisation of emergency mobility will already attend to the quality of both emergency and governance as mobile and composed of complex combinations of quite fluid actors. It means attending to emergency governance at the level of plans, mechanisms and mobile practices at various scales. And even if it seeks to decentre a notion of emergency away from an Agambenian notion of 'exception', it should not exclude its possibility. The rest of this section builds on this conceptualisation. It outlines seven key characteristics that future mobilities research could pursue by building on existing and developing concepts and promising areas of enquiry. Through these different facets of emergency mobility, it helps add more flesh to a theory of emergency mobility.

Anticipation

It is clear that various forms of anticipation come to constitute the way mobilities are governed, during and in advance of emergency: as a way to imagine emergency, to render it knowable or graspable before it occurs through risk matrices, scenarios and other imaginative-performative techniques like role-play and exercises that simulate and play out complex mobilities of actors and processes (Lakoff 2007). As O'Grady (2014) shows, anticipations of risk within the UK fire service make calculations over fire risk and the response time of the fire service to travel there, negotiating local road networks in order to arrive promptly at the scene. These journeys may be subjected to increased forms of scrutiny through registration and tracking, and there are just as many occasions when these administrative practices fail.

Within the field of Geography, we don't have to look very far to realise the post-war complicity of academic geographers in the anticipatory practices of cold-war evacuation planning, as Barnes and Farish (2006) have outlined the role of the discipline in the deployment of operational research methods to pursue the evacuation planning of Bremerton, Washington in the mid-1950s. Their models and simulations – a 'refined unreality of abstraction' – helped to refine actual emergency plans following a nuclear attack (Barnes and Farish 2006, 820).

Examinations of emergency mobility are beginning to attend to more complex entanglements of these anticipations and how they render mobility governable in the future, right now. Such perspectives may shed helpful insight into contemporary legislation, such as the Civil Contingencies Act (2004) in the UK, which has helped initiate the wide-scale institutionalisation of emergency and scenario planning for numerous mobility emergencies, from crashes and infrastructural failure, to city evacuation plans and widespread flooding (Medd and Marvin 2005).

Forms of anticipation may shape our apprehension of emergencies. For example, O'Grady (2014) explores specific anticipatory technologies devised to govern fire risk and potential emergencies by anticipating the spread of fires with the complex circulations and mobilities of fire personnel through a system called Fire Service Emergency Cover Toolkit. Displayed in risk maps and matrices, the toolkit is a calculation of probabilities determined by speculation over the type of fire, and the speed of the mobility of fire personnel travelling distances to the site of the fire. For Grady, underneath these models rest fundamental assumptions about the ability to govern an emergency like a fire, such as 'limits to the capacity of emergency responders to become mobile' that ultimately ties together 'the immobility of such security agents' with the 'lives and populations they seek to secure' (O'Grady 2014, 524).

The recognition of models, simulations and scenarios would mark an important move in how we understand the anticipation of emergency mobilities through mediating technologies, with all their attendant politics of expert knowledges, accessibility and representation. There is, moreover, an imperative to dampen any potential claims that these approaches could serve to determine future mobile events. Some authors have described the practice of imaginative scenario planning to be 'less a process of authoritative writing or inscribing predictable outcomes upon a blank page, and more a tentative testing and feeling' (Adey and Anderson 2012, 109). It may not be quite known how the mobile and contingent entities of what appeared to be a simple flood event and the necessity to evacuate local residents will unfold, even within a scenario exercise written to prepare the training of leaders for local emergency response.

Mobilities researchers may be well poised to explore the possibilities of experimental intervention into these forms of anticipatory practice. For instance, Whatmore and Boucher (1993) have detailed the ways in which the formal anticipatory models of a flooding emergency produced by government agencies could be subjected to collaborative reasoning and deliberation by a community interest group, enabling different 'political opportunities and associations' to open up in the anticipation and planning for future emergency.

Coordination

Alongside anticipation, we should include the means by which the distribution of resources, people and technologies – in order to bring an emergency under control – are coordinated. Monahan's (2007) examination of Intelligent Transportation Systems and their control rooms in the United States, for example, illustrates the convergence or creep of multiple functions and purposes in monitoring traffic and passengers for a combination of emergency, security and efficiency purposes, as well as the technological apparatus of the control room in enabling coordination practices. Emergency mobilities require intense forms of coordination and communication between multitudes of actors. These processes work in tandem with efforts to *know* the emergency. For example, digital informatics are being developed to confront emergencies through crowd sourced data, which, several authors (Büscher 2013) have argued, is raising significant concerns for privacy through information sharing.

Other contemporary research is exploring the emergency coordination of mobility through concentrations of networks and infrastructures by mediated, augmented and situated action (Büscher 2007). Gordon (2012), for example, illustrates how attention to control rooms and the management of emergency mobilities takes us much closer to the notion of mobility as a 'practical accomplishment', requiring labouring people, things, protocols and technologies to work together. The spaces where emergency mobilities are intensely monitored and governed deserve our attention, these are spaces of remarkable concentration of coordination and decision-making such as control rooms and information hubs for organisations like MSF.

Situated research can tell us more about how the governance of mobilities in emergency are highly provisional and involve substantial uncertainties over whether an incident or an emergency has occurred, involving other searching, missing and finding practices, as we will see below. Moreover, as Gordon shows, activities that coordinate mobilities 'are precarious and practical accomplishments whereby agency is delegated and dispersed through the associations it keeps' (2012, 122).

Managing emergency mobilities involves control room processes of sense-making, classifications through formal codings and collaboration with distant actors. In other words, it may involve a whole series of volatile and sometimes partial and uncertain relations and practices to perform them. This demands that we temper our understandings of emergency mobility governance as necessarily successful, and instead focus more on the mobile practices, organisational structures and cultures, through which governance is performed.

Mobile Machines

Fredriksen (2014) has recently conceptualised the movement and construction of emergency humanitarian spaces, which she suggests are made up of complex and mobile assemblies of equipment and other objects and related practices (see also Duffield 2010). It is in this sense that the study of emergency mobilities could attend to the construction of spaces of humanitarian emergency not only through the predominance and stasis of the camp, but through mobile machines, objects and the making of temporary places (Smirl 2008). For example, *Médecins Sans Frontières*' emergency 'kit' is noted to have been based on a mobile medical apparatus developed by the Red Cross in relation to calls for aid to a blitzed city during the Second World War. The result was what Peter Redfield has called 'a mobile template for crisis response around a principle of flexible standardisation' (2008, 157), to the extent that the provision of humanitarian aid in emergencies is made up of the assembly of standardised mobile equipment. Taking the shelter kits and tents supplied by the IFRC to produce an emergency shelter and relief for those made homeless, Fredriksen argues that 'they actively participate in stabilising and (re)ordering spaces as, at least provisionally, *humanitarian* spaces' (Fredriksen 2014, 150).

If we are to account for the movement of mobile technologies in emergency, we should also account for their access which is often highly differentiated. The late Smirl (2015) has identified several issues in what we could term an automobility of emergency humanitarianism. Just as the SUV and its Land Rover predecessor has been understood by various authors to offer a cocoon-like capsular interior of neoliberal citizenship for today's kinetic elites (Campbell 2005; Graham 2011; Mitchell 2005), in emergency humanitarianism, it offers aid workers an enclosure between one secure location to the next. This is exactly what Smirl sees articulated among international aid workers as an 'impenetrability within the vehicle, and of being above the land through which they travel' (Smirl 2015, 43).

The propensity of emergency humanitarian mobilities that are mediated in this way is for insulation from the perceived security threats going on around them. This separation means that they are shuttled between networked enclaves, what Duffield (2010) has described as the bunkers of an 'archipelago' of international space. Aeromobility offers similar distancing qualities. This time Smirl details the aeromobility arrangements for NGOs following the 2004 Indian Ocean Tsunami in Aceh, Indonesia.

> The UN set up a parallel transport system including almost daily flights to and from certain coastal cities (Calang, Meulaboh). Other organizations such as Oxfam, invested in their own helicopters). Still other organizations such as CHF International used, Mission Aviation Fellowship (MAF) a Christian NGO specialising in flying light aircraft in remote locations. The effect was that international staff, many of them visiting experts, consultants and staff from headquarters experienced the post-tsunami space in an extremely fluid, mobile manner [...] reinforc[ing] the divide between international aid workers, arriving to help, and their intended beneficiaries. (Smirl 2014, 123)

To put it more clearly, the spaces of emergency intervention are mobile in such a way that they can be easily distanced geographically, socially and politically from those who need it. And yet, as assemblies of objects, they are not necessarily immutable, even if they are internationally standardised. As Redfield explores, while MSF's mobile apparatus embodied 'the technical principle of modular mobility [...] the kit is ultimately an open container', in other words, 'it remains available for appropriation into a wide range of projects' (Redfield 2008, 165). It may well be desirable to enable a mutability to mobile equipment, so it can be easily adapted with materials and components supplied from local sources. As Fredriksen shows, the IFRC's emergency kit is made 'standardised to be interchangeable' (156).

A theory of emergency mobility should therefore attend to more mobile conceptions of the spaces of humanitarian aid. These mobile spaces of standardised as well as adaptable machines appear to produce distancing effects, and even the 'islanding' of emergency governance and aid, from the communities they are seeking to protect.

Absence

Emergencies produce absences that are generative of mobilities. Necromobilities and geographies of death and the dead are common both after, but also during, emergency. As Jassal (2014, 3) argues, we should be 'questioning the mundane, political and everyday ways in which the state, mourners and industry professionals transport, dispose and handle the materiality of the dead'. The mobile dead, as much as the living, may also exist on systems of records and administrative systems used by emergency services to govern and coordinate according to different thresholds of life.

However, Edkins (2011) has explicated the vast and inadequate bureaucratic systems through which some people slip; these are the absences of the missing, lost from natural disaster, civil war, state terrorism or maybe sheer incompetence. While mobilities researchers should engage more fully with these instruments in the recording and tracking of the alive, dead or the missing of emergency, and the important valuations of those lives in decision-making, the missing also implies important and mobile practices of looking and finding, addressed at the partial or traces of the living and the dead.

During emergency, huge resources can be drawn on to search, find and recover the lost, the dead or the barely living. Yarwood (2012) has shown that emergency recovery entails a whole set of mobilities and accompanying practices of sensing and seeing in order to retrieve and find what has been lost, whether those on an exposed remote landscape, or we could think of the recent Malaysian Airlines flight MH370.

Mobilities of searching are not only technological but can be highly honed, involving learnt skills of detection conducted in partnership with others moving through a landscape. As Yarwood describes his involvement in a Search and Rescue exercise conducted by a mountain rescue team:

> I could see the torch beams of other teams sweeping the moor. Elsewhere, a flashing strobe zoomed across the moor at inhuman speeds, marking the progress of a search dog as it bounded over the moor seeking air-scents of potential casualties to follow. The radio crackled periodically, broadcasting messages from teams and controllers also engaged in the search. (Yarwood 2012, 25)

Finding the missing then may require performed mobilities of trawling over land or sea, perhaps in the case of MH370, which has required enormous efforts to track back through satellite telemetry of distant oceans, photographed at a lower resolution than the imagery of Mars (McNutt 2014). The adaptation and improvisation of known techniques and technologies has been common. The response has meant sifting through highly contested waters and fraught geopolitical boundaries where to move and search is not an unproblematic offer of goodwill, but potentially an active performance of sovereignty (Steinberg 2014).

As with the contingencies of control rooms, the emergency mobilities of finding and searching for those who have gone missing reveals an indeterminacy of mobilities, that could keep on going for an unknown period of time to places not yet thought. It indicates how emergency mobilities may not be certain, they may not leave the heavy footprint, or concrete traces we expect, the trails to be tracked, watched, picked up, stored, counted, and sorted, and yet gone without much of a trace. In this sense, argue Parr and Fyfe (2013), searching is both a 'messy practice, characterised by confusions', whilst the absence produced by what is lost in emergency and must be found acts as an 'active category' in directing resources and practices. The mobilities of emergency governance may be more about the lost, and how the lost is recovered into the known, or never known at all.

Inhuman

As various authors have argued, there has been much political and ethical work to see emergency and other categories of calamity like disaster as socially embedded and constructed, dependent upon inadequate and unequal societies, or structural failings in economies and politics. While these notions have lent critical import to analysis of the inadequacies of mobility policies during emergencies, others suggest that there might be a ceiling limit to our 'ability to order or regulate our transactions' with the world (Clarke 2010, 29). Emergency mobilities seem to be stirred by the irregular interruptions of the earth, weather patterns, fire, cyclones and tsunamis, an exorbitant nature, and just as contingent relations with other very mobile 'life forms and elemental processes' (Clarke 2010, 29), or what Sarah Whatmore has described as when the 'unexamined parts of the material fabric'; when the stuff that constitutes our 'everyday lives becomes molten' (Whatmore 2013, 37). Of course, Law (2006) has shown how the inadequacies of governance met with the exorbitant materialities of foot and mouth disease to effectively produce the 2001 outbreak in the UK.[1]

There is a point here about whether an exorbitant planet of molten and mobile materialities simply exceeds our abilities to plan, prepare or respond to them in emergencies. Clarke also indicates elsewhere that we might value the experimental and improvisory ways that societies have learnt to live and move under threat of emergency and disaster, what Whatmore (2013) calls the 'forcing thought' of the world, without necessary submission to an emergency governance apparatus which seeks to overdetermine the response:

> they are equally occasions which oblige human populations to respond with experiments of their own. Many of those peoples who still live in relatively close proximity to the rhythms and upheavals of the Earth have learned how best to shelter from extreme events, when to move to safer ground, how to channel excess energies, what to cache or stockpile, and when to fight fire with fire. (Clarke 2014, 32)

Perhaps, then, it should not be an afterthought for us to consider how emergency mobilities may exceed attempts at governing them in a manner which is not always about the allocation of fault or guilt, but a realisation of the 'force of all manner of earthly powers – bodies, codes, devices, models, documents and proteins among them' (Whatmore 1993, 36).

Let us take the outbreak of cholera in Haiti following the 2010 earthquake; that three years after the earthquake, half a million people remained without potable water, adequate shelter or modern sanitation. The conditions helped exacerbate the spread of a cholera epidemic. The outbreak itself, however, was found to have been introduced by UN peace keepers carrying the cholera bacteria from Nepal (Montalo et al. 2013). For Sheller, this reminds us that 'diseases too make use of vectors of mobility with no respect for the borders of states or islands, bodies or cells' (Sheller 2013, 199).

From many perspectives, an example such as this reminds us to juxtapose the power geometries of mobilities of those caught up in emergency and those attempting to solve it. Current legal action contests the UN's biosecurity screening of the Nepalese forces and the negligence of a private contractor to install proper sewage infrastructure within the UN camp at Mirebelais, enabling the spread of the disease (The Guardian 2014). Indeed, the mobilities of what has been called a geopolitics of celebrity of senior politicians and elites – a political vulturism – anxious to witness the scene, further opens up the unevenness of this geometry of access or capability to move. The scene Sheller (2013) urges us to consider in Haiti was the curious hypermobility of those who performed global humanitarian aid, in the face of the radical immobility of the Haitian population forced to endure stark living conditions for the ease of 'ensuring smooth logistical operations'. On the other hand, the case reveals the movements of in-human mobilities of disease and bacteria piggybacking on human and animal bodies, that escaped the notice of governance or were deliberately ignored by it for a contractor's profit.

If a theory of emergency mobility seeks to level critical insight into the practices of those responsible for it, could these complications of inhuman life obscure attempts at identifying negligence? Almost certainly, although the UN is relying upon other factors for its immunity. It also reminds us, even if the Haiti disaster may not fit this mould, that some emergencies may be of such magnitude and complexity, constituted by so many mobilities and contingencies, that their prevention and response may always be inadequate to the task.

Difference

We should be minded that emergency mobilities can reinforce existing geographies of gender and exacerbate them. As mentioned above, the social science response to Hurricane Katrina revealed, through the failed evacuation of the city, and the eventual rehousing of the populace, that emergency mobilities are highly unequal and inequitable (Klein and Smith 2008; Smith 2006). Along racial, ethnic, class and gender lines, research has shown how emergency mobilities reinforce, intensify and produce new and uneven stratifications. However, gender also cuts across other categories of social difference. Recent research illustrates that greater proportions of women are more likely to be living in poverty, make up a higher proportion of the elderly, and share greater responsibility for dependents or relations, all factors limiting their mobility capabilities during emergency. Even more alarmingly, the mobilities of emergency displacement has exposed the increase of gender-based violence from domestic abuse to sexual assault, as research in Haiti and New Orleans has recently demonstrated (Anastario, Shehab, and Lawry 2009; Enarson 1998; Henrici 2010). Mortality rates among women are also greatly increased during emergency than in men (Neumayer and Plümper 2007).

These discussions need situating within the socio-economic context and a politics of place, which force us to remap emergencies mobilities and their governance across the entangled meanings of place, home and domesticity. For example, Morrice (2014) has shown in the context of the 2011 Christchurch earthquake in New Zealand and the Queensland floods in Australia, movements of return can be as traumatic as the process of leaving. Samuels' (2012) fascinating exploration of gendered mobilities following the emergency of an Indonesian neighbourhood, forced to relocate to a purpose-built accommodation block in a secluded village following the 2004 tsunami, illustrates such issues. As Samuels shows, the mobility capabilities of women and their access to jobs and services were disproportionately effected by their rehousing, as compared to men. This can be understood with regard to their new reliance upon expensive public transport and limited private transport to access opportunities in the nearby city. Further investigations may focus on the perception of vulnerability to or in emergency. As Keenan (2014) has recently argued in a study of gendered perceptions of vulnerability to terrorism on transportation infrastructures, 'a person's body – and the societal and individual's expectations about that body – influence the experience of both the place and the experience of vulnerability' (369).

Emergency mobilities tend to be highly plural, journeys that are undergone with others. This withness can flummox efforts to govern them. More often than not, the apparatus and techniques intended to manage mobilities fail to recognise this essential character, separating communities and family groups, or provisioning for only particular kinds of mobile subject. This was seen perhaps most clearly during Katrina when the access of individuals and families to personal mobility technologies, such as cars, was wildly overestimated and optimistic.

What these distributional differences add up to, for several authors, is a politics of mobility that should be more carefully attuned to mobility justice, to the identification and achievement of far more equal mobility capabilities for all, 'to meet their own basic needs' (Sheller 2013, 195).

Times

Emergency mobilities could be characterised by speed – in advance of or during – emergency politics seems to produce the conditions for governing mobilities with the least amount of deliberation possible. As Anderson (2012) has argued, an extra-legislative 'state of emergency' is not the usual governmental mode of response to contemporary emergency, instead, what is exceptional is not necessarily the speed at which constitutional rights are overridden, but 'the speed of response to a situation where life and death were at stake', a situation where timely action is needed.

Just as contemporary emergencies are marked by an onus on fast mobility and fast decisions – 'rapid response' can be found across a range of emergency forms. As Anderson claims:

> in social work teams primed for early intervention in the ordinary crises of daily life; in rapid response Urban Search and Rescue Teams set up to enter disasters zones; or in the Rapid Response Facilities set up to provide 'rapid mobilisation funding' to humanitarian organisations in response to a 'rapid onset disaster'. At a time when welfare is becoming a matter of emergency relief, we also find 'rapid response' in relation to the provision of food or finance or shelter to those in times of personnel crisis. Scaling up, the imperative to respond rapidly is behind the range of emergency measures (Emergency Stabilisations Funds and so on) designed to inject liquidity into the global financial system in the midst of the current Financial crisis. (Anderson 2012)

The speed of emergency is still problematic though. Speed might close out dissenting voices and other forms of response because decisions move too fast. The distribution of aid to the locations where it is determined it is needed may deny the ability to debate or contest the nature, dimensions or response to emergency. The urgent politics of emergency may mean fast deliberations are made when decisions over life, death and the value of life are to be made. Simply put, in the velocities of emergency mobilities, things may have moved on already. Resources, people, aid, may need to move before proper deliberation can take place.

In other contexts, the speed of decisions under emergency mobility may bring a halt to the ordinary running of things. Within New Orleans once more, the absence of legal professionals to ensure due process became a key issue because of the evacuation of most legal officials from the city (Crouch 2013), which was combined with a suspension to the time limit of the writ of *habeas corpus* under the governor's executive powers, keeping prisoners under remand without trial on what was deemed 'Katrina time'. In this sense, the governance of emergency mobilities should alert us to the complex times and temporalities within and outside of the law (Cooper 2014; Opitz and Tellmann 2014). Following Katrina, the legal redress is still ongoing. Because of the complications of federal, state, county and even international law, the homeless who were turned back attempting to cross over a bridge into another county (the town of Gretna) by an armed police blockade have been unable to seek justice even now (Crouch 2013). Despite the apparent right to movement enshrined in the United States Constitution (Cresswell 2001), the constitution actually protects 'inter-state' rather than 'intra-state' travel.

Even if a theory of emergency mobility governance needs to take better account of the enmeshing of law within other contradictory scales that have protected the pejorative and knee-jerk reactions of some officials during exceptional times (protecting a bridge from city residents made homeless), some examples provide more hope for the 'slowing down' of deliberation. As Whatmore and Boucher (1993), following Latour and Stengers, advance, it may be possible to generate spaces of emergency deliberation where political reasoning over the governance of emergencies can be slowed to a point where decisions are done differently. A theory of emergency mobility should be open to these possibilities.

Conclusion

Mobilities, we have seen, are embroiled in the nature and governance of emergency. The modest impetus of this paper has been to attempt to trigger more sustained attention within the field to emergency mobilities, their governance and how we might theorise them. Greater analytical purchase may be possible to interrogate these processes not as an afterthought of other more critical issues within emergency and disaster, to see mobility as simply an outcome of greater forces, nor to simply embrace more nuanced conceptualisations of emergency and governance, often focused upon the spatial thresholds that characterise this work, such as the camp or the prison. In bridging these approaches, the paper has argued that mobilities may in fact be the emergency and come to constitute enhanced and important procedures of governance, with real implications for people's life chances.

Adding flesh to this concept, we then discussed seven particular characteristics common to emergency mobilities. These involve sets of anticipatory practices; processes and protocols of coordination between disparate networks of actors; mobile machines that enable mobile governance to take place; a recognition of the absences that drive many emergency mobility practices of searching, looking and sensing, even for the most faint signatures of life; the identification of inhuman and contingent relations; the production of difference through emergency and emergency governance and, finally, an emphasis on the times, or lack of it, for politics. This is not to say that these qualities are not recognisable in other ways that mobility and governance are organised during normal times. The examples we have seen are often adaptations or improvisations of existing doings, infrastructures or they are intensifications of them. Moreover, the threshold between emergency and normality may be really quite blurred, so that these qualities are recognisable in practices seeking to prepare, prevent or act in advance of emergency.

This is a concept of emergency mobility, however, that has and should emphasise its fragility. Existing attention to the governance of mobilities in emergency has certainly contained a critical edge, able to apportion blame and critique to structural forces, as much as to the key individuals, institutions or choices we have encountered. This has, crucially, been a story of governance, modes of power brought to bear on, and indeed, as emergency and mobility. Moreover, we are used to contemplating the mobilities of emergency as failures, and subject to the demand of just how they could have been done differently or better, especially in view of the identification and achievement of 'mobilities justice'. At the same time, however, far more tentative and contingent relations have been emphasised in the accomplishment of mobility and governance.

We have seen research beginning to account for a less than human take on relations by emphasising the violent or subtle earthly powers of emergency, as much as the complex and uncertain relations between unpredictable mobile life, lives and things. This implies that a concept of emergency mobility should be tempered with notions of mobility and indeed governance as tentative, differently reasoned and highly provisional accomplishments, involving unpredictable and even microscopic mobile actors. This may mean the apportionment of cause, blame or the establishment of equality of mobility capability is frustrated. As we have seen, the governance of emergency mobilities is not at all straightforward, involving messy, fragile, searching, sensing practices, presences and absences, in nearby or remote locations, with non and inhuman life and mediators.

Given these contingencies, it shows us that even as we depart notions of emergency as exception and towards the normalisation of the term within other routine forms of governing emergency and mobility, the kind of exclusions, inequalities and political foreclosures we have been discussing may not be so automatic either. Perhaps, following Clarke (2014) and Honig (2013), these notions invite us to reimagine emergency mobility towards not ever-increasing closures, but providing the potential openings of 'rights, law, hope and politics' (2013, xv).

Acknowledgment

Thanks are owed to the guidance of the editors of the special issue and two anonymous referees. The author also acknowledges the support of a Philip Leverhulme Prize.

Disclosure statement

No potential conflict of interest was reported by the author.

Note

1. Combined with other events near the time, this emergency or, for Law, 'disaster', would lead to the biggest shake-up of UK emergency legislation since the Second World War.

References

Adey, P., and B. Anderson. 2011. "Event and Anticipation: UK Civil Contingencies and the Space-times of Decision." *Environment and Planning A* 43: 2878–2899.

Adey, P., and B. Anderson. 2012. "Anticipating Emergencies: Technologies of Preparedness and the Matter of Security." *Security Dialogue* 43: 99–117.

Agamben, G. 1998. *State of Exception*. Chicago, IL: University of Chicago Press.

Amoore, L. 2006. "Biometric Borders: Governing Mobilities in the War on Terror." *Political Geography* 25 (3): 336–351.

Anastario, M., N. Shehab, and L. Lawry. 2009. "Increased Gender-based Violence among Women Internally Displaced in Mississippi 2 Years Post-Hurricane Katrina." *Disaster Medicine and Public Health Preparedness* 3 (1): 18–26.

Anderson, B. 2012. Rapid Response. *Berfrois*. http://www.berfrois.com/2012/08/ben-anderson-emergency-quick/.

Anderson, B., and P. Adey. 2012. "Governing Events and Life: 'Emergency' in UK Civil Contingencies." *Political Geography* 31: 24–33.

Aradau, C., and R. van Munster. 2012. "The Time/Space of Preparedness: Anticipating the 'Next Terrorist Attack'." *Space and Culture* 15: 98–109.

Barnes, T. J., and M. Farish. 2006. "Between Regions: Science, Militarism, and American Geography from World War to Cold War." *Annals of the Association of American Geographers* 96 (4): 807–826.

Bartling, H. 2006. "Suburbia, Mobility, and Urban Calamities." *Space and Culture* 9 (1): 60–62.

BBC News Online. 2014. "Liberia Declares State of Emergency over Ebola Virus." http://www.bbc.com/news/world-28684561.

Birtchnell, T., and M. Büscher. 2010. "Stranded: An Eruption of Disruption." *Mobilities* 6 (1): 1–9.

Budd, L., S. Griggs, D. Howarth, and S. Ison. 2010. "A Fiasco of Volcanic Proportions? Eyjafjallajökull and the Closure of European Airspace." *Mobilities* 6 (1): 31–40.

Büscher, M. 2007. Interaction in Motion: Embodied Conduct and Movement in Emergency Teamwork. In *Proceedings of the 2nd Congress of the International Society for Gesture Studies: Interacting Bodies*, edited by L. Mondada, 15–18. Lyon: ISGS.

Büscher, M. 2013. "A New Manhattan Project?: Interoperability and Ethics in Emergency Response Systems of Systems." *ISCRAM 2013*, Baden-Baden.

Calhoun, C. 2010. The Idea of Emergency: Humanitarian Action and Global (Dis)Order. In *Contemporary States of Emergency: The Politics of Military and Humanitarian Interventions*, edited by D. Fassin and M. Pandolfi 29–58. Cambridge, MA: Zone Books.

Campbell, D. 2005. "The Biopolitics of Security: Oil, Empire, and the Sports Utility Vehicle." *American Quarterly* 57 (3): 943–972.

Clarke, N. 2010. *Inhuman Nature: Sociable Life on a Dynamic Planet*. London: Sage.

Clarke, N. 2014. "Geo-politics and the Disaster of the Anthropocene." In *Disaster Politics: Materials, Experiments, Preparedness*, edited by M. TIroni, I. Rodriguez-Giralt, and M. Guggenheim, 19–37. London: Wiley-Blackwell.

Collier, S. J., and A. Lakoff. 2008. Distributed Preparedness: The Spatial Logic of Domestic Security in the United States. *Environment and Planning. D, Society and Space* 26 (1): 7–28.

Cook, N., and D. Butz. 2015. "Mobility Justice in the Context of Disaster." *Mobilities* 1–20. doi:10.1080/17450101.2015.1047613

Cooper, M. 2014. "The Theology of Emergency: Welfare Reform, US Foreign Aid and the Faith-based Initiative." *Theory, Culture & Society*, 32(2): 53–77.

Cresswell, T. 2001. "The Production of Mobilities." *New Formations* 43: 11–25.

Cresswell, T. 2008. "Understanding Mobility Holistically." In *The Ethics of Mobilities: Rethinking Place, Exclusion, Freedom and Environment*, edited by S. Bergmann and T. Sager, 129–140. Aldershot: Ashgate.

Crouch, M. P. 2013. "In the Aftermath: Responsibility and Professionalism in the Wake of Disaster." *South Carolina Law Review* 65: 465–503.

Diken, B., and C. B. Laustsen. 2005. *The Culture of Exception: Sociology Facing the Camp*. London: Routledge.

Duffield, M. 2010. Risk-management and the Fortified Aid Compound: Everyday Life in Post-interventionary Society. *Journal of Intervention and Statebuilding* 4 (December): 1–19.

Edkins, J. 2000. "Sovereign Power, Zones of Indistinction, and the Camp." *Alternatives* 25 (1): 3–25.

Edkins, J. 2011. *Missing: Persons and Politics*. New York: Cornell University Press.

Ek, R. 2006. "Giorgio Agamben and the Spatialities of the Camp: An Introduction." *Geografiska Annaler: Series B, Human Geography* 88 (4): 363–386.

Enarson, E. 1998. "Through Women's Eyes: A Gendered Research Agenda for Disaster Social Science." *Disasters* 22 (2): 157–173.

Fassin, D., and M. Pandolfi. 2010. *Contemporary States of Emergency: The Politics of Military and Humanitarian Interventions*. New York: Zone Books.

Fredriksen, A. 2014. "Emergency Shelter Topologies: Locating Humanitarian Space in Mobile and Material Practice." *Environment and Planning D: Society and Space* 32 (1): 147–162.

Foucault, M. 2003. *Society Must Be Defended: Lectures at the Collège de France, 1975–76*. Translated by D. Macey. London: Penguin Books.

Gordon, R. 2012. *Ordering Networks: Motorways and the Work of Managing Disruption*. Durham: Durham University.

Graham, S. 2005. "Cities Under Siege: Katrina and the Politics of Metropolitan America." *Social Science Research Council: Understanding Katrina: Perspectives from the Social Sciences*. http://forums.ssrc.org/understandingkatrina/author/sgraham/.

Graham, S. 2009. *Disrupted Cities: When Infrastructure Fails*. New York: Taylor & Francis.

Graham, S. 2011. *Cities under Siege: The New Military Urbanism*. London: Verso Books.

Grove, K. J. 2013. "From Emergency Management to Managing Emergence: A Genealogy of Disaster Management in Jamaica." *Annals of the Association of American Geographers* 103 (3): 570–588.

Hannam, K., M. Sheller, and J. Urry. 2006. "Editorial: Mobilities, Immobilities and Moorings." *Mobilities* 1 (1): 1–22.

Henrici, J. 2010. "A Gendered Response to Disaster: In the Aftermath of Haiti's Earthquake." *Anthropology News* 51 (7): 5.

Honig, B. 2013. *Emergency Politics; Paradox, Law, Democracy*. Princeton, NJ: Princeton University Press.

Ikeya, N. 2003. "Practical Management of Mobility: The Case of the Emergency Medical System." *Environment and Planning a* 35 (9): 1547–1564.

Jassal, L. K. 2014. "Necromobilities: The Multi-sited Geographies of Death and Disposal in a Mobile World." *Mobilities* 10 (3): 1–24.

Keenan, K. 2014. "Place Ontologies and a New Mobilities Paradigm for Understanding Awareness of Vulnerability to Terrorism in American Cities." *Urban Geography* 35 (3): 357–377.

Klein, N., and N. Smith. 2008. The Shock Doctrine: A Discussion. *Environment and Planning. D, Society and Space* 26 (4): 582.

LA Times. 2014. "Liberia Imposes Ebola Quarantine and Curfew in a Monrovia Slum." http://www.latimes.com/world/africa/la-fg-africa-ebola-liberia-curfew-20140819-story.html.

Lakoff, A. 2007. "Preparing for the Next Emergency." *Public Culture* 19 (2): 247–271.

Law, J. 2006. "Disaster in Agriculture: Or Foot and Mouth Mobilities." *Environment and Planning A* 38 (2): 227–239.

Little, S. 2006. "Twin Towers and Amoy Gardens: Mobilities, Risks and Choices." In *Mobile Technologies of the City*, edited by M. Sheller and J. Urry, 121–133. London: Routledge.

McNutt, M. 2014. "The Hunt for MH370." *Science* 344 (6187): 947.

Medd, W., and S. Marvin. 2005. "From the Politics of Urgency to the Governance of Preparedness: A Research Agenda on Urban Vulnerability." *Journal of Contingencies and Crisis Management* 13 (2): 44–49.

Mitchell, D. 2005. "The SUV Model of Citizenship: Floating Bubbles, Buffer Zones, and the Rise of the 'Purely Atomic' Individual." *Political Geography* 24 (1): 77–100.

Monahan, T. 2007. "'War Rooms' of the Street: Surveillance Practices in Transportation Control Centers." *The Communication Review* 10 (4): 367–389.

Montalo, F., P. Gurian, M. Piasecki, and M. Sheller. 2013. *Post-earthquake Haitian Infrastructure Reconstruction: A Focus on Water & Sanitation in the Town of Leoganetle*. Philadelphia, PA, 1–77. https://mcenterdrexel.files.wordpress.com/2013/04/final-haiti-report.pdf.

Morrice, S. 2014. *Returning "Home"? Emotional Geographies of the Disaster-displaced in Brisbane and Christchurch*. Egham: Royal Holloway University of London.
Neocleous, M. 2006. "The Problem with Normality: Taking Exception to 'Permanent Emergency.'" *Alternatives: Global, Local, Political* 31 (2): 191–213.
Neumayer, E., and T. Plümper. 2007. "The Gendered Nature of Natural Disasters: The Impact of Catastrophic Events on the Gender Gap in Life Expectancy, 1981–2002." *Annals of the Association of American Geographers* 97: 551–566.
O'Grady, N. 2014. "Securing Circulation through Mobility: Milieu and Emergency Response in the British Fire and Rescue Service." *Mobilities* 9 (4): 512–527.
O'Regan, M. 2010. "On the Edge of Chaos: European Aviation and Disrupted Mobilities." *Mobilities* 6 (1): 21–30.
Opitz, S., and U. Tellmann. 2014. "Future Emergencies: Temporal Politics in Law and Economy." *Theory, Culture & Society* 32 (2): 107–129 .
Parr, H., and N. Fyfe. 2013. "Missing Geographies." *Progress in Human Geography* 37 (5): 615–638.
Redfield, P. 2008. "Vital Mobility and the Humanitarian Kit." In *Biosecurity Interventions: Global Health and Security in Question*, edited by A. Lakoff and S. J. Collier, 147–171. New York: Columbia University Press.
Salter, M. B. 2008. "When the Exception Becomes the Rule: Borders, Sovereignty, and Citizenship." *Citizenship Studies* 12 (4): 365–380.
Samuels, A. 2012. "Moving from Great Love: Gendered Mobilities in a Post-Tsunami Relocation Neighborhood in Aceh, Indonesia." *International Journal of Urban and Regional Research* 36: 742–756.
Sheller, M. 2013. "The Islanding Effect: Post-disaster Mobility Systems and Humanitarian Logistics in Haiti." *Cultural Geographies* 20 (2): 185–204.
Smirl, L. 2008. "Building the Other, Constructing Ourselves: Spatial Dimensions of International Humanitarian Response." *International Political Sociology* 2: 236–253.
Smirl, L. 2015. *Spaces of Aid: How Cars, Compounds and Hotels Shape Humanitarianism*. London: Zed Books.
Smith, N. 2006. "There's No Such Thing as a Natural Disaster." *Understanding Katrina: Perspectives from the Social Sciences 11*. http://forums.ssrc.org/understandingkatrina/theres-no-such-thing-as-a-natural-disaster/.
Steinberg, P. 2014. "The Malaysian Airlines Tragedy and South China Sea Geopolitics." *Rhulgeopolitics*. https://rhulgeopolitics.wordpress.com/2014/03/12/the-malaysian-airlines-tragedy-and-south-china-sea-geopolitics/.
The Guardian. 2014. "Haitians Launch New Lawsuit against UN over Thousands of Cholera Deaths." http://www.theguardian.com/world/2014/mar/11/haiti-cholera.
Whatmore, S. 2013. "Earthly Powers and Affective Environments: An Ontological Politics of Flood Risk." *Theory, Culture and Society* 30 (7–8): 33–50.
Whatmore, S., and S. Boucher. 1993. "Bargaining with Nature: The Discourse and Practice of 'Environmental Planning Gain'." *Transactions of the Institute of British Geographers* 18 (2): 166–178.
Yarwood, R. 2012. "One Moor Night: Emergencies, Training and Rural Space." *Area* 44: 22–28.

Re-Assembling (Aero)mobilities: Perspectives beyond the West

WEIQIANG LIN*,**

*Department of Geography and Planning, University of Toronto, Toronto, Canada
**Department of Geography, National University of Singapore, Singapore, Singapore

ABSTRACT *This paper advocates a need to expand mobilities research beyond the West. Employing aviation in Singapore as an example, it demonstrates how the assembling of (aero)mobilities in different contexts never yields passive replicas, but, rather, iterations that develop with reference to one another. This mutual assembling is furthermore a political process, with certain 'global' paradigms being more influential than others. Without transcending a Western focus, mobilities research risks obscuring the highly relational way movement is practically (re)assembled, through complex processes of diffusion, adaptation and re-production across space.*

Introduction

For over a decade, the mobility turn has ushered in new academic insight on the ways social lives are being constructed on the move, disenchanting a long-held belief in the stability and 'place-ness' of societies (Urry 2000). This negation of territorial stasis does not only draw inspiration from the intensifying flows of people transgressing all sorts of borders, but is also informed, at a more minute scale, by the banal motions that permeate everyday life, such as walking, commuting and other routine circulations (Adey 2010a). Beyond people's movements, mobility has also come to encompass the travels of a myriad of things, including internationally shipped goods, information and even planning policies (Graham 2005; McCann 2011; Cowen 2014). Without these flows and relations, society and its life-worlds would purportedly cease to function, let alone be sustained in any coherent way.

Notwithstanding the productive way scholars have highlighted the centrality of mobility, current research has tended to depend heavily on Anglo-American and Scandinavian (loosely defined here as 'Western') experiences for analytical perspective. If unqualified, such empirical biases risk connoting a universality in the way the world – including so-called 'developed' worlds such as 'Europe' – moves, despite clear differences in history, attitudes, technological wherewithal, and policy-related latitudes and constraints across space. Consider, for instance, the comparatively more modest views on transport 'modernity' in the Global South (Wood 2014), or China's expanding car culture amid global fears of climate change (Bradsher 2004). These examples reveal that mobilities, and their meanings, (mis) uses and milieus, often take on quite different significances in different contexts. What is more, how this difference arises is also of theoretical pertinence: not simply as self-evident 'cultural' typologies, but as effects of complex, if uneven, global interactions that engender and compel differentiated ways of moving (Steele and Lin 2014). To the extent that mobilities are thus variegated, current research also needs to attend more closely to these 'other' manifestations, and interrogate how they may emerge in conjunction with the mobilities scholars are already writing about. More than 'a way of having a relation to … the world' (Adey 2010a, xviii), mobilities are, themselves, assembled through relations *in* a much bigger world, marked by diffusions, mutual impacts and political arrangements.

This paper seeks to advance the mobilities literature by turning to these inter-spatial variations and connections. By espousing a more global view on the assembling of mobilities, it argues that present work can be furthered and nuanced through foregrounding and comparing a wider range of asymmetrically constituted mobile iterations, especially in the 'non-West'. I illustrate this by focusing on aeromobilities as my metonymic example. Though an extraordinary type of movement that not everyone partakes in, air travel has attracted 'surging and unprecedented interest' (Cwerner 2009, 2) recently, not least because the 'air and atmosphere' are increasingly regarded as 'zones in which a range of important disciplinary concerns might be addressed' (McCormack 2009, 27). Besides implicating a variety of domains, including culture, technology and politics (Butler 2001; Adey 2010b), air transport is also salient for its internationality, entailing that its activation and appearance can neither be wholly bounded within, nor replicated out of any one context. I illuminate an instance of its distributed formation through Singapore, an Asian city-state deeply entrenched in aerial networks, yet also distanced from the creation of aviation's orders. Resisting the notion that the presence of international rules and standards simply breeds uniformity in the industry, I elucidate how aeromobilities are, in this context, a distinctive culmination of possibilities and constraints that are derived from uneven relations.

The rest of this paper is divided into four sections. In section two, I review some key themes within the aeromobilities literature, and highlight their near-exclusive allegiance to (a particular) 'Western' experience. The next section then considers how assemblage theory can inform a more multiplicitous understanding of mobilities and their hierarchical arrangements globally. After providing a brief context on Singapore, the fourth section elucidates how aeromobilities are always enfolded in slightly different ways in/by the city-state, as distinctive sets of problems (and solutions) arise on the regulatory, technological and geopolitical fronts. I conclude in the fifth section by reiterating the need to attend to mobilities beyond the West, not simply to exhaust 'other' means of making move, but to contemplate how movement's multiple iterations and assemblages might potentially be relational and political outcomes of complex interactions across space.

'Pinning Down' Aeromobilities

Literature on aviation in the social sciences has blossomed in recent years, seeing scholars unearthing and explicating diverse aspects of 'being airborne' (e.g. Millward 2007; Pirie 2009; Adey 2010b). Shunning the utilitarian focus of 'traditional' transport geography – a sub-discipline criticised for its outmoded positivist frameworks – much of this work seeks to re-discover air transport as a meaningfully constituted form of mobility, intricately interwoven with human society, its intents, and its particular methods of organising air travel. As opposed to a staid fixation on hub-and-spoke networks, spatial patterns and regulatory policies (Graham 1995; Oum and Yu 2000), there is now a greater appreciation of the practices, experiences and politics surrounding aviation (Adey, Budd, and Hubbard 2007). It is through such a re-orientation that the phenomenon of air travel can finally be accounted for on its own terms, 'as an object or a culture in its own right' (Adey and Lin 2014, 61).

A full account of this culture is probably impractical in this short space, but a couple of recurring themes are noteworthy. The first concerns the social life of air travel, and the fresh outlooks that it brings. As Adey (2010b, 8) writes, 'aeromobilities have shaped and defined the scope of our movements ...', and 'quite new forms of life ... have been produced by the aeroplane'. These may pertain to the novel sensations, etiquettes and cultures that flight had acquainted (a select segment of European) society with at the beginning of the air age (DeLyser 2010; Budd 2011; Caprotti 2011a); or, in more contemporary times, to the increasingly prevalent lifestyle of global travel, jet-lags, transits and – to throw in the mix for non-elite passengers – draconian airport and border surveillance, that has come to define what a modern, airborne citizen should know, and be accustomed to (Gottdiener 2001; Adey 2004; Bissell 2015). This does not mean that there is equal access to such experiences, as air travel remains a resolutely 'white' space with little tolerance for subjects like stowaways and undocumented migrants (Sheller 2010; Martin 2011). But despite these exclusions, aviation has radically transformed society's expectations and orientations towards the world and others. This unprecedented cultural paradigm has prompted current research to emphasise and finesse what Adey (2010b) terms 'aerial life'.

Related to this 'new' life, scholars are equally interested in how aerial orders are constituted and established. More than giving a superficial run-down of aviation's make-up, they have carefully considered how humans, machines and other constituents need to be precisely aligned and coupled to render air transport operable (Urry 2007, 141). From air ticketing to check-in to baggage and cargo handling, mobilities researchers have underscored how aviation behaves as a complex system, which increasingly relies on computer technologies to regulate its functions and routines (Dodge and Kitchin 2004; Graham 2005; Peters 2009). Beyond facilitation on the ground, Budd and Adey's (2009) work further highlights how software is similarly an integral part of modern piloting, as aircraft monitoring systems take over as the vicarious means by which pilots 'see', and navigate through the sky (see also Adey 2010b; Williams 2011a). Extending from this technical realm are yet other legal and operational elements, including air rights, airways, control zones and air traffic control procedures, which just as crucially order the conduct of aeromobilities. Without a clockwork-like synching of these parts, the air transport system would become susceptible to stalls and disruptions, laying bare society's fragile dependence on aeromobilities (Birtchnell and Büscher 2011).

Aeromobilities research has clearly made much progress in promoting a more robust understanding of air travel. In line with the tenors of the mobility turn, it has productively attended to its subject as less a utilitarian means of conveyance, than a culture in its own right. Yet, like the rest of the field, this corpus has been curiously dependent on Western – or, more precisely, Anglo-American and Scandinavian – experience for insight, despite its focus on a transport mode that is highly globalised. While concentrating on the West is not, in itself, problematic, and has, indeed, been generative of many lessons, I argue that a near-exclusive reliance on it could work against a research agenda meant to avoid pinning down social facts to particular territories and enclaves. As Bhimull (2011) and Caprotti's (2011b) rare work on Caribbean and Fascist Italian air transport briefly alludes, 'alternative' iterations of aeromobility not only do exist; they also commonly develop in conjunction with a variety of contexts, including the dominant West. Insofar as these 'other' emergences and their dynamics remain largely obscured, future work could benefit from a more thorough investigation on them, in affording a more cosmopolitan view on aeromobilities, and how they unevenly materialise across space.

Re-Assembling (Aero)mobilities

Researching mobilities beyond the field's original 'birthplace' in the West – or, in terms of academic practice, in parallel Anglo-American and Scandinavian universities – has, in fact, been a growing call of late, precisely because of the novel theoretical perspectives it portends. As Cresswell (2014, 714) writes in a recent stocktaking piece, 'mobilities research needs to happen outside the UK, Scandinavia and North America' more frequently, for moving – either by air or otherwise – can be 'a very different proposition' elsewhere. This sentiment is shared by Steele and Lin (2014) in a later editorial in *Transfers*, in which they seek to qualify the relationships linking different mobilities in different regions, while arguing, more specifically, that twentieth-century 'Asia was (and is) not simply following in the West's footsteps … [but] became "mobile", "modern" societies in their own right'. Such a comparative stance to (re)imagine mobilities as a multi-centred endeavour not only raises awareness of the diverse mobile realities that exist in tandem. It also impels recognition that mobilities do not, in themselves, have an internal logic that can be pinpointed, and then universalised from the West.

In this context, a deeper appreciation of the tenets of assemblage theory – a paradigm by which (aero)mobilities research is already often enunciated (Budd and Adey 2009; Williams 2010a) – can help to make more concrete sense of this diversity. Expressed as the antithesis of pre-determined wholes, De Landa (2006, 10) defines assemblage as constellations characterised 'by relations of exteriority', where component parts autonomously enter into relations with one another. Anderson and McFarlane (2011, 126) more plainly explain it as such, writing that '[a]ssemblages always "claim" a territory … [b]ut this can only ever be a provisional process: relations may change, new elements may enter, alliances may be broken, new conjunctions may be fostered'. This idea of emergence is not new to mobilities, but most studies tend to ascribe this potential for re-territorialisation to singular contexts, rather than to assemblages that may (re)emerge as revised iterations *elsewhere* (Lagendijk and Boertjes 2013). In allowing formations to relationally spawn and re-generate themselves at new sites, assemblage theory can inspire a greater sense of multiplicity in the realisation of mobilities – not just as transformations over time, but as different forms co-existing across space. Succinctly, it entertains the possibility of 'otherness' within these 'topologically complex and less certain' processes of assembling (Hetherington and Law 2000, 129), exceeding the current practice of tracing and retracing the same, familiar paths of inquiry within the West.

Mobilities' multiplication across space is, however, not random, but closely intertwined with the asymmetric relations that order them. As De Landa (2006, 19) avers, populations of assemblages do not simply appear as assortments of estranged emergences, but are 'synthesised' to form 'larger-scale assemblages' of particular hierarchical arrangements. Ong and Collier's (2005, 4) concept of 'global assemblage' presents an excellent example of such hierarchies, stressing that the articulation of individual systems is rarely independent, but filtered through certain paradigmatic 'global forms' or dominant templates. While, in some respects, the consequent production of a range of worldwide phenomena such as the 'nation-state' and 'economy' (and, I add, aeromobilities) appears homogenising, global assemblages and their specific expressions, nonetheless, do not so much result in uniformity as further entrench a politics of difference, as templates get imposed and activated in different – and differently positioned – milieus. Here, the agency of local actors in receipt of these global forms becomes key, as they selectively adopt, adapt and/or circumvent such impositions to create new formations (McFarlane 2009, 562). By highlighting how assemblages are thus contentiously negotiated and re-made, assemblage theory can uncover another relational geography in mobilities, this time, composed of 'templates' and (revised) 'copies' produced at the interstices of dominance and subordination.

These insights gained from brushing mobilities up against the nuances of assemblage theory return us to the need to expand mobilities research beyond its current preoccupation with particular Western examples. By highlighting the complexity of the assembling process *across space*, these perspectives do not just call for a focus on diversity, but also valorise the political dynamics by which mobile assemblages become juxtaposed and differentiated in relations to one another. Indeed, analogous to how spaces – from cities (McCann 2011) to territories (Allen and Cochrane 2007) – are currently conceptualised, mobilities can likewise be vested with such a relational quality: of being enfolded through elements and influences both 'here' and 'there', 'near' and 'far' (Allen 2011). At times, this may see dominant forms constraining the possibilities of assembling in some places; at others, it may mean that the same templates double as 'solutions' that can be adopted/adapted to answer local (mobility) needs. Certainly, there are instances of movement – especially quotidian ones such as walking – that remain untouched by such distributed processes. But without taking the first step to acknowledge that mobilities may have a more global (and globally intertwined) face than what current scholarship propounds, it is easy to mistake and de-politicise the experiences of the West as the sole, and falsely bounded, model for a universal knowledge.

Singapore's Aeromobilities

To elaborate on these ideas, I turn to Singapore to illustrate how aeromobilities are thus contentiously assembled through interactions and negotiations that link the Southeast Asian city-state to the world. As a former British trading colony, Singapore has long placed international transport at the heart of its economic strategies, particularly after its independence in 1965 when it was politically severed from its hinterland, Malaysia (Lin 2015). Within this policy, the aviation sector has been particularly emphasised and heavily invested on by the Singaporean state. In 1981 for instance, a brand-new airport, Changi, was constructed at a hefty cost of S$1.5 billion (US$700 million in 1981), to replace a previous, capacity-limited aerodrome. Against competition from regional rivals willing to become Southeast Asia's primary air hub, the country became one of the earliest champions of a liberal aviation policy; (briefly) served as the sole non-Western operator of the Concorde; and boasted a top national carrier repeatedly voted first in international surveys (Raguraman 1986, 1997). These achievements are not simply driven by a desire to be aerially developed like the West. Instead, they are a reflection of the particular circumstances under which the city-state became tied in a race for aeromobile pre-eminence in its part of the world.

Despite these aspirations, Singapore never really possessed the clout commensurable with its development, constituting the other side of the story of this context. The city-state is neither an originator of aviation science, nor does it manufacture aircraft. This already excludes it from making any direct changes to, or innovations in, the way aviation is being technologically realised. Expertise aside, Singapore is moreover a young nation, whose autonomous history does not even go back far enough to give it a voice in the regulatory aspects of air transport during the industry's formative years. Much of its pre-1960 aerial development could, instead, be attributed to the British colonial government, which not only represented the settlement (as Malaya) in international forums, but also supplied it with technical personnel, expertise and financial resources. Given this state of affairs, Singapore's aeromobilities must be seen as something never organic to itself. Even to this day, insofar as 'the world's aerial order remains one that is heavily skewed towards the logics of the Western world' (Lin 2014, 97), its aviation development continues to be swayed and constrained by global rules that it is more an adopter/adapter, than an author of.

Such a confluence of global structures, regional rivalries and local ambitions suggests the presence of a distinctive context within which aeromobilities in Singapore thrive. The following will examine how these factors practically affect and revise the articulation of three fundamental 'component parts' of aviation, namely, its regulation, technologies and geopolitical exercise. Textual analyses of over 300 Singaporean aviation industry newsletters, official public releases, archived government documents and International Civil Aviation Organisation (ICAO) meeting reports spanning from 1984 to 2014 will inform most of this discussion. Besides these texts, evidence is also triangulated through 27 interviews with Singaporean aviation officers (comprising 15 with air hub planners and 12 with technical personnel), as well as existing literatures on civil aviation in Singapore. Together, these data shed light on the limitations Singapore faces as a practitioner of aeromobilities in Southeast Asia. They put in context the adaptive strategies the city-state has undertaken to carve out a sphere of influence for itself, in a congested world of often larger and historically more pre-eminent aviation players.

On Route Regulation

As a maturing industry, civil aviation may seem to have been organised into fixed regulatory regimes that are universally applied. Yet, though major conundrums concerning nations' use of the air were settled post-Second World War (Budd 2009), this appearance of 'stability' (Williams 2011b, 254) remains something of a mistaken identity. Indeed, the spread of global aviation orders beyond their origins in North America and (some parts of) Europe is not unproblematic, as they are not all compatible with individual countries' interests. One rule that often proves contentious pertains to the 'Freedoms of the Air', a route-trading system established in the 1940s, after the 1946 Bermuda I Agreement between the US and the UK (Little 1949). Since its institution, the mandatory practice of having states negotiate bilateral, or, more recently, multilateral, treaties before commercial air services can commence has dictated the (non)freedom of routes, stopover possibilities, frequencies and capacities of international flights for the last seventy years (Button and Taylor 2000). Turning to Singapore can reveal how this 'global assemblage' (Ong and Collier 2005), originally designed to protect the aviation industries of post-war Western countries, is received in places that did not author it. Equally crucially, performing such a contextual shift can underline the adaptive strategies places like Singapore have had to assume to (re) assemble a different, more favourable outcome for themselves.

Though bound by this international convention, Singapore has consistently distanced itself from the protectionist ideals espoused by 'Freedoms of the Air'. As historicised in a newsletter its civil aviation authority published:

> Quickly while the bombs were still dropping, men [*sic*] frantically [tried] to come to agreement among themselves on some order to govern the air … Men [*sic*] learned the war lesson well. They extended the [air's] jealous protection to the economics of the Air. The Agreement from the 1944 Conference, therefore, left a gap for men [*sic*] to hassle, namely the economic rights to carry passengers, cargo or mail. (Aviation Views 1986a, 6)

For Singapore, the requirement for rights trading before carriage forms a 'gap' created out of (certain, Western countries') war-time anxieties that its administrators now have 'to hassle'. Rather than protecting *its* interests, this regulatory assemblage, in fact, disadvantages it. Specifically, Singapore's small size connotes that it has 'a low market potential in terms of generation of air transport demand', leading to 'few airlines [being] attracted to Singapore, … [and, thus], weak bargaining strength in getting rights' (Raguraman 1986, 58). A former negotiator drove home the gravity of this to Singapore's carriers by calling air rights 'the lifeblood of an airline', without which 'there is no point in having an airline' (Lim 2000, vol. 2). Constrained by an international restriction inimical to its geographical circumstances, Singapore cannot passively defer to customary rights-trading practices prevalent around the globe, but requires additional labours to strengthen its aerial position in the world.

To do so, authorities have applied themselves to be proactive intermediaries in the procurement of air rights, becoming the (human) agents that would lubricate the formation of favourable trading arrangements and assemblages for aviation to flourish in. Working alongside instituted rules, but also seeking to overcome them, these officers are responsible for holding regular consultations with political counterparts to maintain close bilateral ties, and to explore new avenues for treaties, or air services agreements (ASAs), at every opportunity. As one interviewed officer puts it, 'ASAs don't … come about … just because [Singapore wants] to sign [one]. There's also other work like … to network, to cultivate, to build good relations with other countries'. Without a viably attractive market to fall back on, the city-state became dependent on such diplomatic relations for its air links to basically function, rendering them not so much protective, but, rather, *constitutive* of aeromobility itself. By making network-building a conscientious part of aviation, a 'new' mobile regime has contextually emerged here, predicated upon maintaining goodwill with counterparts, and a humanly performed kind of regulatory (re)assembling (cf. Budd 2009).

A 'civil aviation promotion team' was additionally set up 'to actively encourage and persuade foreign airlines to operate services to Singapore' (Aviation Views 1986b, 6). According to another interviewee, this has the benefit of activating air rights that are already secured (by launching actual flights), and reducing the likelihood of foreign governments declining future rights expansion (on account that 'our [Singapore's] carriers are flying, but their carriers are not'). Here, Singapore introduces another round of human-level inputs to its aviation space, involving members of the promotion team actively marketing the city's hub to foreign carriers, and supporting their operations in Singapore through (overseas) passenger outreach and promotional partnerships. By engaging in these unusual forms of business development on behalf of *foreign* airlines, the aim is to accelerate the establishment of air links with Singapore, and compensate for its disadvantage within the dominant route-trading system. This is another example of how the small, Asian city-state alternatively assembles its aeromobilities, not by accepting a ready format, but by adding creative strategies and new managerial functions to an unfavourable global arrangement.

These revisionary tactics may sometimes attract further protectionist measures from foreign governments, necessitating additional rounds of adaptation. Particularly, as the success of Singapore's airline-and-air-hub model consolidated, the city-state became a target of even stricter regulatory sanctions. Such curtailments were especially pronounced before the 1990s, when its flagship carrier, Singapore Airlines (SIA), was periodically subject to North American and West European authorities' harassment and revocation of rights (Outlook 1981a). More systematically, Australia implemented, in 1979, an international civil aviation policy (ICAP) that prevented SIA from offering (cheaper) 'point-to-point' fares between Australia and the UK via its Singapore hub, allowing Qantas and British Airways to undercut it in price. In reaction, Singapore rallied similarly affected neighbouring states to lobby against the ICAP, culminating in the policy's repeal in 1981 (Outlook 1981b). Like the softer persuasions it normally engages in, Singapore's more overt protest here dispels any illusions that aviation is activated, and defended, the same way universally. They intimate the complex and asymmetric dynamics taking place when a global assemblage, originating in the West, is applied outside the West, precipitating a relational politics of regulatory freedom and benefit, and, correspondingly, new methods in re-assembling and re-making (aero)mobilities, elsewhere.

On Technology

International rules on what constitutes approved technological knowledge and expertise in aviation present another set of impositions that can lead to differentiated outlooks and conduits in realising aeromobilities. While it is true that most modern jet aircraft follow largely similar operating principles (precisely because they are manufactured by only a few approved suppliers in North America and Western Europe, unless one flies a Russian or Chinese aeroplane), unequal influence over these principles can often mean the difference between technological dominance and dependency (Engel 2007). Further splintering outcomes, local factors such as the availability of technical experts, access to financial resources, regional traffic requirements, and even material considerations like the terrain (or waterscape) over which aircraft operate, can render the actualisation of flight a highly disparate experience from context to context. Without attending to these nuances, the technological practices and arrangements cited in existing literatures potentially tell only half the story – or perhaps the depoliticised stories of their (Western) inventors – as to how aeromobilities are expertly assembled. In this respect, the Singapore example can, again, afford an alternative understanding, by speaking from a position that cannot take such expertise for granted.

An apposite case that can demonstrate the intertwined technological relationships Singapore shares with other places in assemblage is found in its efforts to augment airspace capacity. Faced with rapidly growing regional air traffic, Singapore is pressured to find remedies to its saturated airspace, and to use technology to enhance its carrying capacity. But, to do so, it is, again, bound by globally standardised air traffic management rules that not only dictate the number of aircraft it can support travelling in and out of its airports, but also determine how any capacity increases are to be achieved (ICAO 2007). While these regulations are not more stringently applied to Singapore than other jurisdictions, they nevertheless oblige it to seek solutions that the aforementioned rules prescribe (Lin 2014). Having no expertise of its own, neither alternative solutions by which it can contest the existing assemblage, the production of aeromobilities in Singapore is encountered, here, not as a set of ready logics or inventions, but, rather, a collection of requirements that it has to begin assembling through learning. This time, it must labour to plug itself *into* the global assemblage of aviation expertise, rather than repudiate it, by virtue of its technological dependence on others.

Its undertaking, in the mid-2000s, to increase airspace capacity over the South China Sea exactly exemplifies this slightly more convoluted pathway to attaining technological 'breakthrough'. The project involved tailoring the US-pioneered RNP-10 (required navigation performance) standard to suit two regional airways, which were to see a reduction of legally permissible longitudinal separation between successive aircraft from 80 to 50 nautical miles (i.e. a 60% capacity increase as a result of the reduced spacing) (ICAO 2008). Devoid of its own technologists – or what an interviewee describes as 'great mathematicians' – Singapore had to turn to American expertise to help it implement this foreign standard. Expressly, it procured the assistance of Washington DC-based aerospace company, CSSI, to perform the necessary collision probability assessments, and to survey regional equipment for their compatibility with the planned reduction (ICAO 2008). Insofar as this external help enabled the 'proper' accounting of risk, the anticipatory logic (Budd and Adey 2009), and by extension, the credibility, of Singapore's intensified airspace was thereby maintained. In a relational act of assembling involving global prescriptions necessitating global assistance, this is an example of how a logic of dependency works to link iterations of aeromobilities both 'here' and 'there' (Allen 2011), exacting a power dynamics that divides the technological 'haves' from the 'have-nots'.

Singapore's importation of these paradigmatic technological assemblages proves to be an even more complex undertaking when one adopts a regional view. To reap the benefits of reduced aircraft separation, all contiguous airspaces around Singapore must conform to the city-state's new standards, in order for reduction limits to be harmonised along entire airways. Apart from RNP-10, a second round of reductions based on (similarly imported) Automatic Dependent Surveillance-Broadcast (ADS-B) technology would deepen these integrative relations. To transpose ADS-B – originally used for mountainous navigation in Alaska – to the South China Sea, Singapore persuaded neighbouring states to (prematurely) enrol in technological projects that began as its own. While Indonesia and Vietnam had to earmark three strategically located islands for the installation of (land-based) signal receivers, the entire region was obliged to equip its aircraft with ADS-B capability, to continue using the capacity-augmented airspace (ICAO 2014). Besides having to import frameworks from the US, Singapore thus had to assemble a regional (air)space of technological cooperation too, complete with its own political dynamics and material considerations tied to the South China Sea. A 'new' logic of expertise appropriation distinguished from, but simultaneously connected to, the West and its currently pre-eminent global technological assemblage was, as such, at work here, entailing a selective, two-step diffusion of standards from one (dominant) assemblage to multiple others.

These examples do not so much negate the 'technogeopolitics' (Butler 2001) observed between (some) North American and European contexts as extend it in a different guise in Southeast Asia. On the one hand, it is ostensible that, for technologically less adept players like Singapore, aeromobilities are not simply conjured out of a ready armature of home-grown technological inventions, but rather require the technical assistance of dominant countries like the US; on the other hand, these mobilities are not wholesale replicas of paradigmatic models (McCann 2011), being often re-constituted forms that have undergone further (re)assembling at the local and regional levels. Significantly, these complex dynamics engender not one, but multiple, re-configured circuits of technological dominance and dependency, projecting a much more intricate technological reality in (aero)mobilities than one solely involving Western invention and (self-)appropriation. Indeed, no single context can exhaust the richness of these technological cultures, which demand a ceaseless search for their cross-border relations, mutual dependencies and strategic adoptions/adaptations across a variety of connected terrains and (global) assemblages.

On Geopolitics

A final insight that a view from Singapore is able to afford pertains to the distinctive ways in which aeromobile geopolitics is practised and constitutively (re)assembled outside the West. While it is uncommon for states lacking the same aerial might as Western (super)powers, such as the US and UK, to engage in overt forms of power demonstration (excepting perhaps China and Russia), air 'wars' are nonetheless fought among secondary state-players, if involving other interests, using other means, and having other implications. Yet, these scuffles are not completely isolated from the aforementioned superpower geopolitics either, but tied to histories of Western colonialism, present-day neo-colonial practices and discourses, and land-air disputes rooted in a Westphalian system of states, sovereignty and territorial integrity (Ó Tuathail 1996). In this section, I project these concerns up to Southeast Asian skies, and consider how Singapore 'battles' with its neighbours for airspace control, in another round of aeromobility assembling that is a subset of global geopolitics.

The world's airspace must foremost be recognised as being divided into many discrete, but contiguous, zones of aerial control known as flight information regions (FIRs). As a means of spatial organisation for navigation, FIRs are not the same as sovereign airspaces, which refer to states' legal ownership of air columns directly above their territories. While FIR control is usually delegated to the state whose sovereign territory underlies the airspace in question (ICAO 1974), the process of determining air traffic services responsibilities in a myriad of airspaces that can span international high seas, areas of uncertain sovereignty, and territories deemed too close to each other[1] intimates an opening for dispute. Most notably, some states possess the authority to provide air traffic services in FIRs that span across multiple sovereign jurisdictions. Because of these disjunctive spatial configurations, the control of airspace is often a potential sticking point that states fight over.

In times of ambiguity, a global discourse advocating the delegation of navigation control to the technologically superior guides international policy. For most North American and European powers, this is not a considerable issue, given that their technological lead grants them undisputed oversight of disproportionately large FIRs, including over the Atlantic, Indian and Pacific Oceans. But for others, this logic (conversely) accords them with more tenuous rights. Such was the experience of Singapore in the 1970s and 1980s, when Malaysia formulated plans to obtain control of the airspace British administrators had handed down to Singapore. Opposed that the latter was directing air traffic above its territories, as well as between its land masses (i.e. in the South China Sea), Malaysia proposed at the first ICAO Asia-Pacific Regional Air Navigation Meeting in 1973 that it should assume responsibility of nearly the entire FIR instead, from the Straits of Malacca to eastern South China Sea (Figure 1) (ICAO 1973). This proposal threatened Singapore's aeromobilities, as it would deprive the city-state of a viable terminal airspace. Without the same credentials as technologically advanced states, Singapore was thus susceptible to a different kind of contest little known to the West. Linking these divergent experiences was a geopolitical logic that as much ordered the world's aeromobilities, as privileged particular (dominant) states in aerial security.

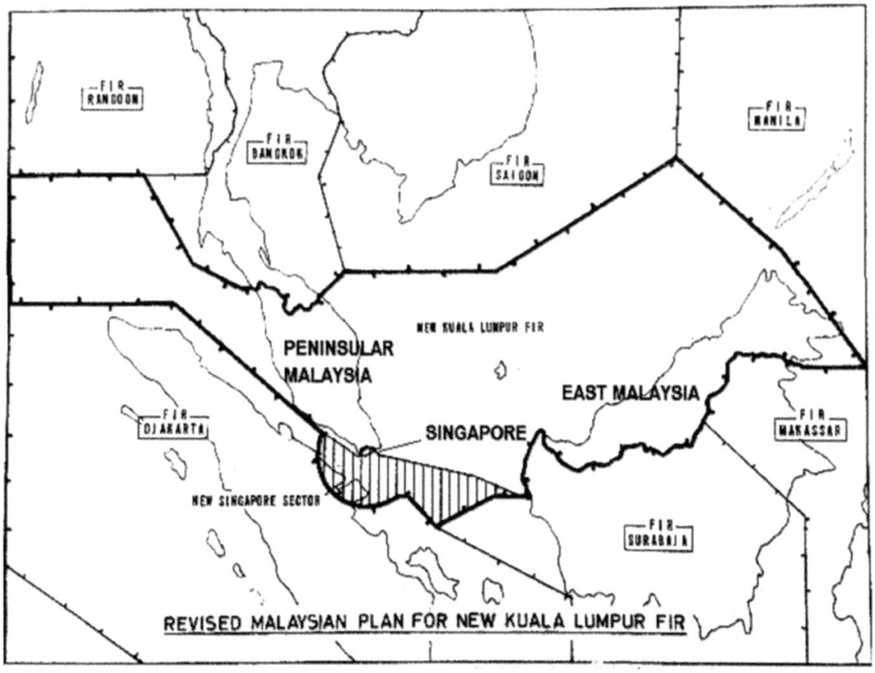

Figure 1. Map showing Malaysia's 1973 plan to partition Singapore FIR (in bold).
Source: national archives, UK, FCO 76/671.

Notwithstanding, while this discourse excluded Singapore from the privilege of making definitive claims, it also provided it with a rubric to counter Malaysia's contests. Following the 1973 meeting, Malaysia had succeeded in acquiring control over the overland portions of the FIR (coinciding with Peninsular and East Malaysia), but not over the South China Sea (ICAO 1973). To pre-empt future challenges to its remaining airspace, Singapore strategically assembled a set of geopolitical practices that would see it enjoin itself to the navigation interests of technological leaders. Consider its statement at the 22nd Assembly in 1977:

> My country has been implementing, and will continue to implement, the standards and recommendations laid down by ICAO for the safety of air navigation, because we believe that [only so can] further progress … be made … (ICAO 1977, 68–69)

Pledging to conform to ICAO's 'standards and recommendations' signalled Singapore's willingness to align itself with the technological programmes of leading (Western) states. At the 24th ICAO Assembly, Singapore further elaborated that.

> We have relentlessly upgraded and improved our aviation facilities to meet [convention] demands … The new Singapore Changi Airport and the computerized Long Range Radar System which we have installed to provide a high standard of air traffic services for flights through the Singapore FIRs are but two recent examples of Singapore's commitment … (ICAO 1983, 44)

This 'commitment' to an FIR of 'a high standard', using state-of-the-art equipment, represented Singapore's attempt to neutralise its regionally weaker position by performing a global 'politics of representation' (Sharp 2011, 271). Here, the city-state not only aligned itself with a network of decision-makers with greater clout than itself, but also invoked a familiar tongue of technological superiority to justify its continued reign over airspace. In relating local/regional concerns to a global reasoning amenable to its cause, it managed to build for itself a reputation of (mimicked) technological competency, by which it has also since deterred any would-be contenders.

Aeromobility, as expressed through its geopolitics, again, exhibits a very different complexion when taken outside the West. In this instance, Singapore did not so much resort to brute force or any new defining logics, as opportunistically re-packaged an existing global assemblage of leading geopolitical reasoning and practice to suit its local/regional challenges. Specifically, in wading through troublesome 'relations of exteriority' (De Landa 2006) emerging vis-à-vis Malaysia, Singapore's response to the global Westphalian system of FIRs was to capitalise on it, and to (re)assemble – both discursively and materially – a set of clout-amassing resources that would help it preserve its aerial stake within that order. A turn to other contexts than those of the usual Western air powers is able to accent these less prominent, though no less salient, air wars fought on the sidelines. It demonstrates that the 'rest of the world' is not reticent about the geopolitics of aeromobilities, only re-constituting it in less recognisable ways, for different purposes.

Conclusions

Echoing Cresswell's (2014, 714) call for mobilities research to take place 'outside the UK, Scandinavia and North America', this paper has advocated, through the example of aviation, the need to assume a more extended view of how movement acquires new valences and meanings in different contexts. Although it is true that most expertise on the 'new' mobilities paradigm has thus far – and very richly so – been informed by Western experiences, it is improbable that a universal knowledge can be exhaustively inferred by consulting the same sphere over and over again. This paper has considered three fundamental aspects of aeromobilities – namely their route regulation, technological arrangement, and geopolitical exercise – in relations to Singapore, as a demonstration of the situational departures that are possible when the view is cast beyond the West. In each case, a re-constituted outlook, sensibility and method of 'making move' inevitably emerges, suggesting that further theoretical insights can be gleaned from such a re-orientation. To bring (aero)mobilities research to the next plane, it is imperative to reflect more seriously on what happens when mobilities meet the road 'elsewhere', for, often, most of the world is not straightforwardly mobile along with the West.

This is, however, not an unthinking push for diversity for its own sake. As an intellectual manoeuvre interested, too, in how iterations of movement are invested in one another, extending the research to the non-West can show how mobilities are as much mutually fashioned through constellational relations across space, as they are individually constituted. Drawing on assemblage theory, and, particularly, its attention to *populations* of coalescences (De Landa 2006), this paper has been conscientious to capture how aeromobilities' enfolding in Singapore does not take place in isolation, but is, rather, always imbricated with other formations. This may pertain to the architectures that govern the global organisation of aviation, or to the more proximate regional orders that interfere with Singapore's aeromobile territorialisations. To the extent that connections can be made across different intersecting formations, mobile assemblages must also be recognised for their highly relational and mutually implicated nature, being configured simultaneously 'here' and 'there' (Allen 2011). Attending to this multitude of formations does not only compound the myriad of relations mobilities are entangled in; it also piques questions about how contained previous renderings of mobilities in (parts of) North America and Europe actually are.

The uneven manner in which co-existing assemblages affect one another introduces a further political dimension to (aero)mobilities. Amid their constellational interactions, assemblages are often strictured by uneven power relations that render some formations more influential than others (Lagendijk and Boertjes 2013). As this paper highlights through aviation, such paradigmatic forms, or what Ong and Collier (2005) call 'global assemblages', tend to be monopolised by dominant countries in the West, so much so that aeromobilities in places like Singapore are rarely free-forming, but subject to particular (regulatory, technological and geopolitical) templates. But while these hierarchical arrangements suggest a kind of exclusionary politics impinging on some localities' (aero)mobile autonomy, they are also what gives impetus to divergent ways of contesting and (re)assembling (aero)mobilities, as local agents resist, accommodate and/or capitalise on these global assemblages. Indeed, in Singapore's case, dominant standards have, at times, propelled it to find ways to circumvent unfavourable impositions; and, at others, afforded it with (import-ready) solutions to address local/regional threats. These diffusions necessitate a keen awareness of the complex processes mediating the proliferation of certain mobility forms as paradigms, as well as their latent politics. Rather than absolute 'best practices', they may well be evidence of similarly iniquitous networks that, at best, provide a basis for adaptation, or, at worst, dictate how societies ought to move globally.

This paper submits that moving mobilities research beyond its theoretical and empirical 'birthplace' in the West is of critical importance. As assemblages characterised by inter-spatial relations and emergences, its subject-matter – mobilities – can never be pinned down too permanently in any one territory (Anderson and McFarlane 2011), or sphere of understanding, The examples on Singapore and aviation broached above are clearly only a small step in venturing towards this wider repository of possibilities. By exemplifying an unconventional mode of mobile assembling, and in allowing conversations to take place across contexts, this paper hopes to have at least persuaded researchers that mobilities can hardly be seen as the sole preserve of, or wholly bounded within, 'the UK, Scandinavia and North America'. In an age where movement has enabled global intercourse, it seems only fitting that these 'alternative' modes are duly analysed for their mutual imbrications and significances in future scholarship. Without which, the field could risk the very territorial provincialism it critiques.

Acknowledgements

Special thanks to my two anonymous reviewers and editors, Allison Hui and James Faulconbridge, for their thoughtful feedback, as well as to Tim Cresswell and Pete Adey for having inspired so many of the ideas in this paper. All errors, however, remain the author's.

Disclosure statement

No potential conflict of interest was reported by the author.

Funding

This work was supported by the UK Commonwealth Scholarship Commission [grant number SGCR-2011-72].

Note

1. FIRs do not correspond to territories in these cases as fine aerial divisions would increase pilots' workload and compromise safety.

References

Adey, P. 2004. "Surveillance at the Airport: Surveilling Mobility/Mobilising Surveillance." *Environment and Planning A* 36 (8): 1365–1380.

Adey, P. 2010a. *Mobility*. London: Routledge.
Adey, P. 2010b. *Aerial Life: Spaces, Mobilities, Affects*. Malden, MA: Wiley-Blackwell.
Adey, P., L. Budd, and P. Hubbard. 2007. "Flying Lessons: Exploring the Social and Cultural Geographies of Global Air Travel." *Progress in Human Geography* 31 (6): 773–791.
Adey, P., and W. Lin. 2014. "The Social and Cultural Geographies of Air-transport." In *Geographies of Air Transport*, edited by A. Goetz and L. Budd, 61–71. Farnham: Ashgate.
Allen, J. 2011. "Topological Twists: Power's Shifting Geographies." *Dialogues in Human Geography* 1 (3): 283–298.
Allen, J., and A. Cochrane. 2007. "Beyond the Territorial Fix: Regional Assemblages, Politics and Power." *Regional Studies* 41 (9): 1161–1175.
Anderson, B., and C. McFarlane. 2011. "Assemblage and Geography." *Area* 43 (2): 124–127.
Aviation Views. 1986a. "What on Earth is Freedom of the Air?" *Aviation Views* 2 (3): 6.
Aviation Views. 1986b. "New Promotion Team to Woo New Airlines to Singapore." *Aviation Views* 2 (1): 6.
Bhimull, C. 2011. "Caribbean Airways, 1930–32: A Notable Failure." *Journal of Transport History* 33 (2): 228–242.
Birtchnell, T., and M. Büscher. 2011. "Stranded: An Eruption of Disruption." *Mobilities* 6 (1): 1–9.
Bissell, D. 2015. "Virtual Infrastructures of Habit: The Changing Intensities of Habit through Gracefulness, Restlessness and Clumsiness." *Cultural Geographies* 22 (1): 127–146.
Bradsher, K. 2004. "Banished Bicycles; China's Car Culture Hits Some Potholes." *The New York Times*, January 11. http://www.nytimes.com/2004/01/11/weekinreview/the-world-banished-bicycles-china-s-car-culture-hits-some-potholes.html.
Budd, L. 2009. "Air Craft: Producing UK Airspace." In *Aeromobilities*, edited by S. Cwerner, S. Kesselring, and J. Urry, 115–134. London: Routledge.
Budd, L. 2011. "On Being Aeromobile: Airline Passengers and the Affective Experiences of Flight." *Journal of Transport Geography* 19 (5): 1010–1016.
Budd, L., and P. Adey. 2009. "The Software-simulated Airworld: Anticipatory Code and Affective Aeromobilities." *Environment and Planning A* 41 (6): 1366–1385.
Butler, D. 2001. "Technogeopolitics and the Struggle for Control of World Air Routes, 1910–1928." *Political Geography* 20 (5): 635–658.
Button, K., and S. Taylor. 2000. "International Air Transportation and Economic Development." *Journal of Air Transport Management* 6 (4): 209–222.
Caprotti, F. 2011a. "Visuality, Hybridity, and Colonialism: Imagining Ethiopia through Colonial Aviation, 1935–1940." *Annals of the Association of American Geographers* 101 (2): 380–403.
Caprotti, F. 2011b. "Overcoming Distance and Space Through Technology: Record Aviation Linking Fascist Italy with South America." *Space and Culture* 14 (3): 330–348.
Cowen, D. 2014. *The Deadly Life of Logistics: Mapping Violence in Global Trade*. Minneapolis: University of Minnesota Press.
Cresswell, T. 2014. "Mobilities III: Moving On." *Progress in Human Geography* 38 (5): 712–721.
Cwerner, S. 2009. "Introducing Aeromobilities." In *Aeromobilities*, edited by S. Cwerner, S. Kesselring, and J. Urry, 1–21. London: Routledge.
De Landa, M. 2006. *A New Philosophy of Society: Assemblage Theory and Social Complexity*. London: Continuum.
DeLyser, D. 2010. "Feminism and Mobilities: Crusading for Aviation in the 1920s." In *Geographies of Mobilities: Practices, Spaces, Subjects*, edited by T. Cresswell and P. Merriman, 83–96. Farnham: Ashgate.
Dodge, M., and R. Kitchin. 2004. "Flying through Code/Space: The Real Virtuality of Air Travel." *Environment and Planning A* 36 (2): 195–211.
Engel, J. A. 2007. *Cold War at 30,000 Feet: The Anglo-American Fight for Aviation Supremacy*. Cambridge, MA: Harvard University Press.
Gottdiener, M. 2001. *Life in the Air: Surviving the New Culture of Air Travel*. Lanham, MD: Rowman and Littlefield.
Graham, B. 1995. *Geography and Air Transport*. Chichester: Wiley.
Graham, S. 2005. "Software-sorted Geographies." *Progress in Human Geography* 29 (5): 562–580.
Hetherington, K., and J. Law. 2000. "Guest Editorial: After Networks." *Environment and Planning D: Society and Space* 18 (2): 127–132.
ICAO. 1973. *Report of the First Asia/Pacific Regional Air Navigation Meeting*. Montreal: International Civil Aviation Organisation.

ICAO. 1974. *The Convention on International Civil Aviation: Annexes 1 to 18*. Montreal: International Civil Aviation Organisation.

ICAO. 1977. *Assembly 22nd Session: Minutes of the Plenary Meetings*. Montreal: International Civil Aviation Organisation.

ICAO. 1983. *Assembly 24th Session: Plenary Meetings, Minutes*. Montreal: International Civil Aviation Organisation.

ICAO. 2007. *Procedure for Air Navigation Services: Air Traffic Management*. Montreal: International Civil Aviation Organisation.

ICAO. 2008. "Report of the Second Meeting of ICAO SEA-RNP Implementation Task Force." Accessed July 26, 2012. http://www.icao.or.th/meetings/2008/rnpsea_tf2rpt.pdf

ICAO. 2014. "Report of ADS-B Seminar and Thirteenth Meeting of ADS-B Study and Implementation Task Force." Accessed June 17. http://www.icao.int/APAC/Meetings/2014%20ADSBSITF13/Report%20of%20ADS-B%20SITF13-25th%20V4-Rev.pdf

Lagendijk, A., and S. Boertjes. 2013. "Light Rail: All Change Please! A Post-structural Perspective on the Global Mushrooming of a Transport Concept." *Planning Theory* 12 (3): 290–310.

Lim, C. B. 2000. *Transportation in Singapore* [Accession Number 002358/2]. Singapore: Singapore National Archives.

Lin, W. 2014. "The Politics of Flying: Aeromobile Frictions in a Mobile City." *Journal of Transport Geography* 38: 92–99.

Lin, W. 2015. "Cabin Pressure: Designing Affective Atmospheres in Airline Travel." *Transactions of the Institute of British Geographers* 40 (2): 287–299.

Little, V. 1949. "Control of International Air Transport." *International Organization* 3 (1): 29–40.

Martin, C. 2011. "Desperate Passage: Violent Mobilities and the Politics of Discomfort." *Journal of Transport Geography* 19 (5): 1046–1052.

McCann, E. 2011. "Urban Policy Mobilities and Global Circuits of Knowledge: Toward a Research Agenda." *Annals of the Association of American Geographers* 101 (1): 107–130.

McCormack, D. P. 2009. "Aerostatic Spacing: On Things Becoming Lighter than Air." *Transactions of the Institute of British Geographers* 34 (1): 25–41.

McFarlane, C. 2009. "Translocal Assemblages: Space, Power and Social Movements." *Geoforum* 40 (4): 561–567.

Millward, L. 2007. *Women in British Imperial Airspace: 1922–1937*. Montreal, QC: McGill-Queen's University Press.

Ó Tuathail, G. 1996. *Critical Geopolitics: The Politics of Writing Global Space*. Minneapolis: University of Minnesota Press.

Ong, A., and S. Collier. 2005. "Global Assemblages, Anthropological Problems." In *Global Assemblages: Technology, Politics, and Ethics as Anthropological Problems*, edited by A. Ong and S. Collier, 3–21. Malden, MA: Blackwell Publishing.

Oum, T. H., and C. Yu. 2000. *Shaping Air Transport in Asia-Pacific*. Aldershot: Ashgate.

Outlook. 1981a. "Closed Skies Over Frankfurt." *Outlook: House Newspaper of the SIA Group*, December: 3.

Outlook. 1981b. "ICAP is Dead." *Outlook: House Newspaper of the SIA Group*, June: 1.

Peters, P. 2009. "Airborne on Time." In *Aeromobilities*, edited by S. Cwerner, S. Kesselring, and J. Urry, 159–176. London: Routledge.

Pirie, G. 2009. *Air Empire: British Imperial Civil Aviation, 1919–1939*. Manchester, NH: Manchester University Press.

Raguraman, K. 1986. "Capacity and Route Regulation in International Scheduled Air Transportation: A Case Study of Singapore." *Singapore Journal of Tropical Geography* 7 (1): 53–67.

Raguraman, K. 1997. "Airlines as Instruments for Nation Building and National Identity: Case Study of Malaysia and Singapore." *Journal of Transport Geography* 5 (4): 239–256.

Sharp, J. P. 2011. "Subaltern Geopolitics: Introduction." *Geoforum* 42 (3): 271–273.

Sheller, M. 2010. "Air Mobilities on the US–Caribbean Border: Open Skies and Closed Gates." *The Communication Review* 13 (4): 269–288.

Steele, M. W., and W. Lin. 2014. "Mobilities and Modernities in Asia: History, Historiography, and Be (com)ing on the Move." *Transfers* 4 (3): 43–48.

Urry, J. 2000. *Sociology Beyond Societies: Mobilities for the Twenty-first Century*. London: Routledge.

Urry, J. 2007. *Mobilities*. Cambridge: Polity Press.

Williams, A. J. 2011a. "Enabling Persistent Presence? Performing the Embodied Geopolitics of the Unmanned Aerial Vehicle Assemblage." *Political Geography* 30 (7): 381–390.

Williams, A. J. 2011b. "Reconceptualising Spaces of the Air: Performing the Multiple Spatialities of UK Military Airspaces." *Transactions of the Institute of British Geographers* 36 (2): 253–267.
Wood, A. 2014. "Learning through Policy Tourism: Circulating Bus Rapid Transit from South America to South Africa." *Environment and Planning A* 46 (11): 2654–2669.

The Boundaries of Interdisciplinary Fields: Temporalities Shaping the Past and Future of Dialogue between Migration and Mobilities Research

ALLISON HUI

Department of Sociology, Lancaster University, Lancaster, UK

ABSTRACT *This paper contributes to the interdisciplinary fields of migration and mobilities research by temporalizing understandings of their boundaries – places where differences have been entrenched and some concepts have remained beyond negotiation or dialogue. While the creativity and boundary-crossing potential of interdisciplinary fields is often set in opposition to disciplines, which define and regulate appropriate concepts and knowledge, such characterizations obscure how interdisciplinary fields have boundaries that change over and in relation to time. This paper therefore uses three temporal dynamics – a/synchronicity, sequencing and accumulation over time – to consider the evolving boundaries that have limited collaboration between these fields. By tracing past discussions of concepts such as 'transnationalism', 'mobility' and 'methodological nationalism', it highlights the contingency and complexity of dialogue between these fields, and how they, like disciplines, 'define what it is permissible not to know'. The new concept of 'migrant exceptionalism' is introduced to acknowledge the boundaries created through privileging 'migrants' as unique and continuously relevant subjects. Both migration and mobilities scholars are seen to perpetuate migrant exceptionalism, and countering it through the study of sometimes-migrants is identified as a means of modulating existing boundaries and opening new spaces for interdisciplinary dialogue.*

Dialogue between interdisciplinary research fields involves the mobilities of many concepts which traverse disciplinary boundaries, are applied to varied cases and support new contributions. Flows of ideas between migration and mobilities scholars have, for instance, contributed to understandings of subjectivities, sexuality and dwelling (Conradson and Mckay 2007; Crang and Zhang 2012; Mai and King 2009), the intersecting movement of humans and objects (Basu and Coleman 2008; Burrell 2008, 2011), negotiating multiple returns home (Ahmed et al. 2003; King and Christou 2011), and power dynamics playing out over and through borders (Fortier and Lewis 2006; Gill, Caletrio, and Mason 2011; Richardson 2013). Yet accounts of these flows have often been directed more towards supporting and justifying new knowledge than acknowledging the boundaries, complexity and contingency of interdisciplinary exchange. As Barry, Born, and Weszkalnys (2008, 42) note, 'any analysis of the inventiveness of particular kinds of interdisciplinarity must attend to the specificity of interdisciplinary fields, their genealogies and multiplicity'. This paper therefore looks at the movement of ignored or contested ideas between the 'inherently interdisciplinary' fields of mobilities research and migration studies (Castles 2007; Favell 2001, 397; Hannam, Sheller, and Urry 2006), in order to extend understandings of their interrelated genealogies (Adey et al. 2014; Fortier 2014).

In particular, it contributes to discussions of the interdisciplinary fields of migration and mobilities research by temporalizing understandings of their boundaries – places where differences have been entrenched and some concepts have remained beyond negotiation and dialogue. A considerable amount of research has already considered interdisciplinary exchanges, including the structures and characteristics that support successful careers and collaborations (Boix Mansilla, Dillon, and Middlebrooks 2002; Lyall et al. 2011; National Academies (US) et al. 2005), what leads interdisciplinary experiments to fail (Holland 2014; Strober 2010), distinctions between types of interdisciplinary engagements (Barry et al. 2008), and how dialogue can reveal and challenge disciplinary assumptions (Garrow and Shove 2007). Much of this work, however, has minimally situated interdisciplinary interactions within fields that have negotiated and evolving boundaries, not only possibilities and opportunities.

This stems in part from prevalent narratives that frame the possibilities of interdisciplinarity in relation to the closure of disciplines. In order to understand interdisciplinarity, Abbott (2001, 135) suggests, one must delve into the *Chaos of Disciplines* because 'interdisciplinarity presupposes disciplines'. Tracing the development of social scientific disciplines, Abbott (2001, 130) suggests that they are not only crucial for organizing university structures and labour markets, but also for reproducing and bounding epistemologies – disciplines 'create modes of knowledge that seem, to the participants, uniquely real' and 'define what it is permissible not to know'. Similarly, Barry et al. (2008, 20–21) note how 'disciplines discipline disciples': 'ensuring that certain disciplinary methods and concepts are used rigorously and that undisciplined and undisciplinary objects, methods and concepts are ruled out'. Interdisciplinarity has conversely been positioned as creating space for 'boundary transgressions' and movement beyond 'existing knowledge corpuses' (Barry et al. 2008, 21). Drawing on Dogan and Pahre's work (1990), Urry (2000, 210) suggests that 'innovation results from academic mobility across disciplinary borders, a mobility that generates what they call 'creative marginality'. The diversity of assumptions, precedents, frames and approaches informing interdisciplinary engagement support creative interactions with varied audiences, methods, empirical sites or theoretical precepts. The promise of such interdisciplinary engagement between migration and mobilities researchers has been long noted (Blunt 2007; Hannam, Sheller, and Urry 2006; King 2012).

The potential for creativity and collaboration, however, is only one dynamic, and as Barry et al. (2008) show emphasizing the innovation of interdisciplinarity can hide the varied logics and exchanges at work. As Bourdieu suggests (1984), fields are united by common values and stakes, but also remain sites of struggle. The epistemological commitment of both migration and mobilities research to studying systems of global movements necessitates dialogue between researchers from anthropology, cultural studies, geography, history, sociology and other areas. Yet in part thanks to this dialogue, each field is internally divided – with researchers using varied approaches to study related topics (Adey et al. 2014) and diverse empirical cases obstructing the creation of shared theories (Castles 2010, 1566; King 2012, 138). The internal complexity of interdisciplinary fields is glossed over in critiques that acknowledge only early publications (Kalir 2013) or oversimplified 'founding premises' (Franquesa 2011): these fail to recognize that any consensus is a negotiated, relational and partial accomplishment. Moreover, while some scholars engage across (inter)disciplinary borders, others do not. The potential for mobilities researchers to engage with those studying migration, tourism, transport, communication or socio-technical systems has been taken up, but incompletely and with varied impacts upon these established fields of study. All of these dynamics suggest that there is much more to interdisciplinary fields than is often addressed in summary assessments or compressed histories.

Taking up questions of temporality thus provides a way of looking more closely at how the boundaries of these interdisciplinary fields – migration and mobilities research – have changed, and with what consequences for exchanges and dialogue between them. While Abbott (2001, 150) notes that the relationship between interdisciplinarity and disciplines is long-standing, his assumption that they coexist as complementary forces of intellectual development obstructs recognition of how interdisciplinary fields change over and in relation to time. For Abbott (2001, 134), interdisciplinary interactions are problem driven, and thus have limited life cycles. Yet both migration and mobilities research have examined multiple problems, and neither appears to be losing momentum due to a waning of interest. Indeed, at times they better fit a model of 'academically oriented interdisciplinary research' that seeks to answer new questions than a 'problem focused' one, where the emphasis is upon social or policy relevance (Lyall et al. 2011, ch. 2). Temporal dynamics within the fields, particularly in terms of how concepts are engaged with over and in relation to time, therefore, call for additional attention.

This paper is structured around examinations of the temporalities of circulating concepts – a/synchronous interests in similar concepts; the sequencing of concepts, concerns and audiences; and the accumulation of limited understandings of the social world over longer periods of time. By tracing the movement of concepts that reach limited audiences, receive limited approval and frame limited understandings of the world, the paper highlights the contingency and complexity of interdisciplinary fields, and argues that they can, like disciplines, 'define what it is permissible not to know' (Abbott 2001, 130) and rule out un(inter)disciplinary concepts. Mobilities and migration scholars have not equally embraced the concepts discussed in this paper, and addressing their circulation is a part of both recognizing the limits of potential interdisciplinarity and situating past and future contributions.

In addition, the discussion highlights how attending to the circulation of concepts can be important for reflexive assessments of research. As Bourdieu (1990) notes, the logic of the analyst differs from the practices of those she studies. Reflecting upon the temporalities of conceptual categories and circulations is therefore important for acknowledging and responding to the distortion analysts impose upon understandings of the social world. While interdisciplinary fields may start off as spaces in which to engage with concepts and cases not legitimated by other disciplines, over time they too can develop assumptions and boundaries that require critical reflection.

The paper takes up these themes by addressing the circulation of several different concepts. It first discusses how concepts of transnationalism and mobility have circulated, and with what implications for temporalities of interdisciplinary interaction. It then discusses pairs of concepts that seek to counter each other's limits. After reviewing how 'methodological nationalism' points out the limitations of normalizing and privileging 'nations', it introduces the new concept of 'migrant exceptionalism' to highlight similar limitations with the normalizing and privileging of 'migrants'. Throughout, the temporalities of conceptual mobilities are used to highlight how interdisciplinary fields are shaped by limits and boundaries as much as boundary crossing.

Asynchronous Engagement with Transnationalism

The circulation of ideas upon which research depends, and to which it contributes, is reliant upon particular moments of synchronization. Schedules must be synchronized so that ideas can flow between people during meetings or interviews. The synchronization of research outputs with policy or news cycles is important for finding a wider audience and encouraging the circulation of findings. Within interdisciplinary dialogues, the temporal alignment of interests matters as well. That is, taking the contingency of social life seriously requires an acknowledgement that interdisciplinary dialogues are never inevitable, and that the circulation of ideas between fields can face obstacles related to their temporality. To highlight this point, I turn to a particular concept – transnationalism – and consider how the temporality of its circulation within migration studies was important for subsequent engagement with mobilities research.

Since the 1990s, transnationalism has been 'arguably the dominant paradigm in migration studies' (King and Christou 2011, 454). Though the term 'transnational' began to appear several years earlier, the first major attempt to specify it as a conceptual and analytic framework came from Basch, Glick Schiller, and Szanton Blanc (1994). They define transnationalism as:

> The processes by which immigrants forge and sustain multi-stranded social relations that link together their societies of origin and settlement. We call these processes transnationalism to emphasize that many immigrants today build social fields that cross geographic, cultural and political borders. (Basch et al. 1994, 7)

Though this definition was novel, the practices it highlights were not. Finding a concept to bring them together, however, was important because despite the precedents for these practices:

> What was lacking was a cogent theoretical perspective to illuminate their similarities so that they could be identified as in some sense 'the same'. Absent this perspective, such instances would remain isolated historical tales without cumulating, in any sense, toward the development of new typologies or predictions. (Portes 2003, 875)

Basch et al. (1994, ch. 2), therefore, drew conceptually upon discussions of migration, nations, world systems and race to outline several theoretical premises underlying studies of the transnational. While discussions initially focused on the transnational practices of elites, and migrants from the developing world, calls for explorations of '*middling* forms of transnationalism' have subsequently examined experiences such as gap years and study abroad visits (Conradson and Latham 2005, 229). Within migration research, discussions of transnationalism questioned the naturalization of nation states as contexts for studying social phenomena, as well as the set of concepts that presume or reinforce this frame.

This questioning of nation states, and of assumptions that societies are rooted within their boundaries, has also been a concern within mobilities research (e.g. Urry 2000). Yet this interest was not synchronized with engagement around transnationalism. The transnational turn started several years earlier, with discussions of migration scholars 'turning transnational' (Bailey 2001) and a 2003 special issue highlighting the 'growing acceptance of a transnational perspective among migration scholars' (Levitt et al. 2003, 565). Only after the heralding of the 'new mobilities paradigm' (Sheller and Urry 2006) and introduction of the journal *Mobilities* in 2006 did similar discussion occur as a part of an explicit mobilities turn.

The temporalities of these discussions – of transnationalism, on the one hand, and more-than-national mobilities, on the other – is important for understanding the somewhat muted enthusiasm for mobilities research amongst migration scholars. Despite explicit recognition that 'migration studies are crucial to the field of mobilities research' (Hannam, Sheller, and Urry 2006, 10), the last decade has suggested that engaging with the mobilities turn is not similarly 'crucial' for migration scholars. King (2012, 143) notes this ambivalence: 'on the whole, geographers seem not to have "bought into" the mobilities paradigm in the same way that they have carried out research on migrant transnationalism'. Though the reason for this remains 'something of a mystery' for King (2012, 143), I argue that it in part relates to asynchronous interest in similar concerns. Encountering discussions of transnationalism and mobilities in sequence, rather than simultaneously, could lead the latter to appear as 'more of the same', despite its different ontological positioning (elaborated in the next section). In an academic culture where new contributions are prized and marking one's work as distinctive is crucial, there is little reason to spend time on repetitive conversations. One of the reasons migration research was deemed crucial to mobilities work was that its discussions of transnationalism were extremely pertinent for studying a world of global mobilities. Yet the sequencing of this influence also meant that initial discussions of mobilities were not entirely novel to migration researchers.

What this highlights is that both the timing and direction in which ideas flow matter for how interdisciplinary engagements emerge. While mobilities and migration research have shared concerns about how social relations span societies, and how border crossings and global travel fuel transnational connections, their asynchronous engagement with these concerns limited the extent to which they have been examined as a part of mutually engaged interdisciplinary dialogues.

Sequencing Concerns about Mobility

As the previous section highlighted, the asynchronous circulation of ideas between interdisciplinary communities has consequences for dialogues due to how the sequencing of ideas affects assessments of their contributions. Though in this case the result was largely ambivalence, in other cases the sequencing of ideas can lead to open contestation. This corresponds with different understandings of sequencing as (a) subsequent, emergent intervals in time or (b) strategic ordering and prioritization. In the former, concepts can be seen to have relationships of temporal succession, particularly over longer periods such as years or decades, but these are not taken to be indicative of their relative importance since long-standing concepts can remain incredibly pertinent. When the focus moves to how authors prioritize concepts within an existing vocabulary, however, one concept coming before another can be an indicator of strategic importance and the politics of ordering. To expand upon the latter, I consider understandings of 'mobility' and 'migration' in each of these fields.

Looking at understandings of 'mobility' within migration and mobilities research highlights important ontological differences. Within journals such as the *International Migration Review* or the *Journal of Ethnic and Migration Studies*, there is discussion of 'social mobility' between classes and economic brackets, or 'spatial mobility', such as migrants' inter-city movements. In these cases, the concept remains predominantly focused on people (i.e. migrants), with a particular concern for the link between spatial displacement and changes in social position (e.g. Faist 2013). Mobility is a quality of a person, and can occur in relation to objective space or social hierarchy.

Mobilities research, however, emphasizes that multiple mobilities are entangled in social processes and structures, including 'the large-scale movements of people, objects, capital and information across the world, as well as the more local processes of daily transportation, movement through public space and the travel of material things within everyday life' (Hannam, Sheller, and Urry 2006, 1). This marks a clear difference in ontological focus – mobilities research considers multiple interrelated actors – human (backpackers, car drivers, enthusiasts) and non-human (backpacks, cars, binoculars) (Dant 2004; Hui 2012; Walsh and Tucker 2009). The emphasis upon the movement of things other than humans has been noted as one of the field's contributions to social research (Cresswell 2011, 552) and is important for its aim of countering sedentarist accounts that normalize or privilege stasis (Sheller and Urry 2006; Urry 2007). Paying attention to both humans and objects as a part of larger systems of mobility is key for Urry (2007, 48) because assemblages of mobilities both 'contingently maintain social connections' across space and point to 'the significance of systems that distribute people, activities and objects in and through time-space' (2007, 51). Though not all research within the mobilities turn embraces non-human actors, or discussions of assemblages, the plurality of types of mobilities and units of study is an important part of the field's strategic diversity. By comparison, although migration research incorporates what mobilities scholars would see as different mobilities – of monetary or object remittances, of communications, of ideas and imaginings – these are not often treated as units or actors of equal importance.

For mobilities researchers then, both an interest in (non)human actors and the acknowledgement of multiple mobilities leads to seeing migration as one of multiple phenomena, and one of multiple literatures that contribute to understandings of mobilities. That is, mobility is prioritized as a concept, of which migration is one subtype.

Yet within migration studies, the categories of 'migration' and 'mobility' do not necessarily overlap, let alone have this ordering. Faist (2013, 1642) suggests that in both public debates and academic research, it is increasingly labour migrants, not necessarily welcomed by nation states, who are seen to 'migrate', while highly skilled professionals are not immigrants but 'mobiles'. Castles (2010, 1567) notes the same discursive phenomenon, where 'mobility equaled good, because it was the badge of a modern open society; migration equaled bad because it re-awakened archaic memories of invasion and displacement'. This 'glamorization' of mobile people in public discourse is seen to be problematic, and a reason to question discussions of 'mobile' people within academic circles (Glick Schiller and Salazar 2013, 184).

This concern for the entanglement of public and academic discourses suggests that it is not only the ordering of particular concepts that is at stake – it is also the ordering of audiences. While a concern for public discourse exists in many areas of academic research, migration studies have been strongly engaged in policy-related research due to government concerns about the politicization of immigration (Castles 2010, 1570–1571). In addition, policy-driven consulting has often been a key source of funding for migration research centres (Castles 2010, 1572). For some migration researchers then, the importance of engaging with non-academic audiences means that this differentiation between migration and mobility is one that they must reproduce or to which they must respond. That is, when engaging with non-academic audiences comes first, appropriate vocabularies are negotiated in response.

Another aspect of the debate over 'migration' vs. 'mobility', however, is that prioritizing the latter has been seen as problematic for considerations of social inequality and power. When researchers first engaged in discussions of transnationalism, some critiques focused on its tendency 'to overemphasise the hyper mobility of an elite few' (Collins 2009, 4). Warnings that 'sampling on the dependent variable' could hide the absence of transnationalism therefore led to convergence around the understanding that 'not all immigrants are transnationals' (Portes 2003, 876). In this context, the suggestion that all migrants can be discussed in terms of mobility is contentious because of concerns that this de-emphasizes or erases the inequalities of global migration (Faist 2013; Glick Schiller and Salazar 2013). In an early critique of mobility research, Skeggs (2004, 48) suggests that identifying mobility in 'a horizontal rather than a vertical sense' ends up 'flattening out the differences that have to be crossed' and thus obscuring class, power and the problematic influence of individualism. Even though dimensions of inequality have been addressed within the mobilities turn (Ohnmacht, Maksim, and Bergman 2009), and mobilities research creates space to consider intersectional and systemic aspects of inequality or privilege, for some scholars inequalities and power relations should be more central, and require more specific framings of mobility (e.g. Glick Schiller and Salazar 2013). The politics of what might be lost in prioritizing a general frame of mobilities have thus remained a point of tension between fields.

As this shows, discussions about the concept of 'mobility' and how it relates to 'migration' are intricately connected to political concerns and the contexts of academic practice. The sequence in which concepts are placed can be taken as an indication of the prioritization of particular audiences, actors or politics – whether such ordering was intended or not. Occasional misunderstandings have occurred between these fields – with an interest in 'mobilities' being misread as a lack of interest in inequalities.

What this example further highlights is that for some groups within migration studies, there are impermissible sequences for concepts. That is, some boundaries are not appropriate to cross and some distinctions must not be blurred. As a result, interdisciplinary dialogue with mobilities researchers becomes inappropriate. While some migration researchers have engaged with 'mobility', for others the possibility of doing so has reinforced differences and sparked attempts to regulate what should (not) be known.

What is important, however, is not just the relative positioning attributed to migration and mobility, but the extent to which each concept bounds possible understandings of the social world. Abbott's (2001) discussion highlights how even disciplines are marked by fractal differentiations that rework relationships between concepts over time. It, therefore, becomes important to consider how continued attention to particular concepts can, eventually, accumulate a range of effects with particular limitations and implications for further dialogue.

The Long-term Consequences of Engaging with Limited Concepts

The sequence in which concepts are encountered, or placed, has been highlighted as important for the contingent emergence of interdisciplinary encounters, and enduring frictions within them. But as Abbott (2001) highlights, sequences of interactions also have important consequences for understanding the development of social scientific knowledge over time. He compares the accumulation of understandings about the social world with exploring the streets of a city, suggesting that the particular process through which groups split up and turn down different streets, to explore new territory and learn new things, compounds with implications for how much of the city can be discovered. Over longer periods of time, the paths of many pedestrians can create patterns of investigation that 'systematically ignore places' (2001, 31). That is, depending on the extent to which explorations are concentrated in a particular area, more or less of the city becomes known. While Abbott uses this example to discuss the development of disciplines and subdisciplines, I argue that it is also relevant for thinking about how knowledge accumulates within interdisciplinary communities. If interdisciplinary exchanges are not always problem focused, but are also shaped by shared and contested concepts, then how these affect social explorations is important.

This section therefore looks at two pairs of concepts wherein the circulation of one is initiated to counter the limited paths of exploration enforced by the other. The dynamics between such concepts are important because they seek to destabilize biases that can accumulate over time in relation to what it is permissible to know and not know. They also suggest that there is a danger even within interdisciplinary fields for the entrenchment of core concepts to produce significant gaps in understandings of the social world. To begin outlining these dynamics, we first turn to the pairing of 'nations' and 'methodological nationalism'.

Nations and Methodological Nationalism

As discussed above, the concept of 'nation' has been challenged in discussions of both transnationalism and mobilities. Urry (2000) has questioned the long-standing assumption that nations are natural 'containers' for society, highlighting that social transformations necessitate a rethinking of the role of nations. Wimmer and Glick Schiller (2002, 302) have similarly critiqued what they call 'methodological nationalism' – 'the assumption that the nation/state/society is the natural social and political form of the modern world'. In their analysis, the continuing relevance and persistence of nations as a central frame is tied to academic practices and research processes. National political agendas and research funding, for instance, encourage and constrain researchers – supporting studies that focus on the nation state while making projects based in other countries or spanning across borders more difficult. Wimmer and Glick Schiller acknowledge that there are discontinuities and changing national stances towards immigration, but argue that methodological nationalism has become a pervasive characteristic of migration research. In this context:

> The value of studying transnational communities and migration is not to discover 'something new' – though this represents a highly rewarding strategy of research in our contemporary intellectual environment – but to have contributed to this shift of perspective away from methodological nationalism. (Wimmer and Glick Schiller 2002, 302)

By introducing the concept of 'methodological nationalism', Wimmer and Glick Schiller, whose argument is worth reading in its entirety, point out the limits in knowledge accumulated through a continued focus upon nations. The concept brings attention to how deeply embedded assumptions about the importance of nations systematically ignore some aspects of social life, including transnational processes and experiences.

Challenging the relevance of nations and nationalism is not about denying that they have influence. It is about questioning whether they are always relevant as a primary frame for studying the social world. As Wimmer and Glick Schiller (2002, 326) note, 'every clear conceptual structure necessarily limits the range of possible interpretations, as well as the empirical domains that can be meaningfully interpreted'. The concept of methodological nationalism, and its call to recognize the discontinuous and intermittent relevance of nations, thus acknowledges that multiple interpretations are important for making sense of global societies. By reflexively framing the biases of previous research, it points to the importance of studying areas that have been previously ignored.

Migrant and Migrant Exceptionalism

Within migration studies, the concept of 'methodological nationalism' has circulated widely, contributing in part to the establishment of transnationalism as a popular frame for contemporary research. When considering the interdisciplinary exchanges between migration and mobilities scholars, however, it becomes clear that 'nation' is not the only concept that limits understandings of social life. 'Migrant' is similarly problematic. Bailey (2001, 415) suggests that transnational frameworks for research are marked by a suspicion of how 'migrant types are defined and classified', particularly in relation to nation states, and of 'the conceptual and political usefulness of the migrant types that are produced by the above classification'. A mobilities framework, however, does not stop here. It is also suspicious of how types of mobile people (and flows more generally) are defined and classified. 'Migrants' do not exist a priori, but are linked to nations and countries that 'make' migrants through regulations, visas and border processes. Moreover, 'migrant' is but one of a diverse array of subject-positions relevant for studying mobilities, including tourists, soldiers, drivers, loiterers and the impaired (Adey et al. 2014). The importance or atemporal relevance of 'migrants' cannot be assumed, as it may not be conceptually or politically useful for understanding the complexities of social phenomena.

I therefore argue that the concept of 'migrant' needs to be paired with an understanding of 'migrant exceptionalism' – the assumption that migrants are extraordinary mobile subjects, discrete from other (concurrent) subject positions and central units within methodologies. Though focusing on migrants may be important for specific empirical studies, the accumulated effects of this focus have limits and create absurdities. Halfacree (2012, 213), for instance, suggests that many studies of migration have, despite an attention to movement, privileged settlement in place: since 'any migration is likely to be "temporary" in terms of the duration of a person's life, the very idea of "permanent" migration increasingly seems a product of an implicit assumption of normative sedentarist settlement'. Nonetheless, the dominance of methodological nationalism in post-war research reinforced categories of permanent and temporary migrants, which privileged the isolated moments of border crossing (Meeus 2012, 1777). Subsequently, the temporal frame for migration research has broadened to consider transnationalism and how 'the process of migration does not stop after physical movement to another country' (Meeus 2012, 1779). But if migration does not stop after moving to another country, when does it stop? Or more precisely, when do 'migrant' practices give way to other types of practices, making migration no longer of primary relevance when orienting research?

Too often, this question is given little attention. In part, this is due to the professional institutions, journals, policy frameworks and academic conferences that foreground migration and migrants. Whether the name of a migration research centre, the framing of policies around migrant issues, or conventions of style and terminology in particular journals, the embedding of migrant exceptionalism at multiple levels, in multiple organizations, makes questioning its operation politically sensitive and combatting its prominence a significant challenge. Migrant exceptionalism is no less embedded in social institutions than methodological nationalism. Indeed, it is an offshoot of methodological nationalism, as it depends upon the privileging of a status regulated and enacted by nations. Naming migrant exceptionalism is therefore important for recognizing the consequences and limitations of this assumption, and ensuring that important social dynamics are not being obscured through its reproduction.

Just as with nations, challenging the relevance of migrants is not about denying their existence or importance. It is about questioning the extent to which they are normal or appropriate subjects for research. As Halfacree (2012) highlights, in certain contexts and for specific social or political reasons, it is crucial to maintain distinctions between migration and other phenomena. But before and after migration events, migrants are people – with complex practices and roles. Recognizing this requires not taking the appropriateness of 'migrant' for granted. It is not that 'migrants' needs to be replaced with 'mobiles', but that work is needed to look beyond inherently limited assumptions about 'migrants'.

Though it might appear that migrant exceptionalism is only a problem for migration researchers, this is not the case. Naming and circulating this concept is also important for mobilities researchers. After all, while there is a recognition of the multiple subject positions that (im)mobile people take up, and of the multiple structures that enact such positions, empirical studies have too often focused on only one position or sphere of activity. Studies of cyclists and of air travellers have too rarely considered how one person is often both, at different times. Likewise, literature on migrants and migrant experiences is too seldom cited in discussions of other mobilities. The seemingly normal separation of discussions of cycling, flying and migrating is thus an indicator that even within mobilities research, migrants (and other types of mobile subjects) can be treated as exceptional. It is therefore not enough to suggest as Collins (2009, 18, 2011) does that 'non-migrants' and 'migrants' should be studied in relation to each other, and with a concern for relational mobility and immobility. This still privileges and reproduces categories that are framed from the position of nation states. Within individuals' biographies, 'non-migrant' and 'migrant' roles may be interwoven in surprising ways, or even be largely irrelevant to the practices of everyday life. At the same time that some migrants live in a state of 'permanent temporariness' whereby migration status is a continual obstacle in everyday life (Bailey et al. 2002; Collins 2011), others go through periods when having migrated is of limited relevance. The multiple roles and experiences of diverse and unequally privileged people therefore provide a more appropriate starting point.

While at times migration may seem irrelevant for studying other mobilities, there is much to be gained from further interdisciplinary exchange. Passports, for instance, facilitate entry through physical border crossings, but also act as proof of identity to facilitate access to and communication with financial and bureaucratic institutions. The intertwined practices and politics of migration, finance and digital communication systems could therefore be investigated in more detail. Reflecting upon the limits of migrant exceptionalism could also encourage further consideration of the temporalities of experience at stake in studies of mobilities. Within migration research, maintaining a focus upon migrants has provided vivid understandings of the longer term effects of migration processes upon people's lives – temporalities that have been given limited attention within studies of mobilities.[1] The sequencing of a migration event before other experiences has been shown to affect interactions and mobilities even years later. Useful insights could therefore arise from extending the implications of this approach to people's multiple, sequenced careers as car drivers, migrants, cyclists and smart phone users. In addition, the longer term consequences of migration events suggest the importance of developing mobile methods (Büscher and Urry 2009) that attend to processes and data over longer time periods.

The concept of migrant exceptionalism thus has the potential to encourage new interdisciplinary exchanges because it points to the limitations of how both migration and mobilities researchers have addressed the category of 'migrants'. It does not suggest that migrants must be subordinated to other categories of mobile subjects, but encourages reflection upon their normalization as an exceptional group. In order to counter the limits of migrant exceptionalism, further engagement with cases that embrace multiple subject positions and mobilities systems, with temporally fluctuating prominence, will be important. The next section illustrates two such cases that devote attention to sometimes-migrants.

Exploring Cases of Sometimes-migrants

The first case involves the complexity of pregnant mothers from China who travel to Hong Kong in order to give birth. Since 2001, Hong Kong has seen a considerable increase in the number of Mainland Chinese women who travel there on temporary visas in order to give birth. While in 2001, only 1.3% of the city's live births were to parents who did not have HK permanent residency; by 2011, this figure had grown to 37.4% of the over 95,000 births (Lam and Chan 2013).[2] The surge of birth tourism, which has been supported by niche businesses that aid mothers with entry and find them places to live before their 'emergency' births in the city, was facilitated by two significant events. First, in 2001, a court decision granted Hong Kong Permanent Residency to children born in the city, regardless of the residency of their parents. Then in 2003, the introduction of the Individual Visit Scheme to boost tourism after the SARS outbreak made it possible for Mainland residents to visit Hong Kong without an official tour group (Tsoi 2012). After these developments, birth tourism became an option for both parents hoping to evade China's 'one child policy' and those attracted by the benefits of Hong Kong citizenship, including education and health care provisions (Basten and Verropoulou 2013, 332). After birth, families return to other parts of China. But many leave the city with the anticipation that their child will return – both to apply for documents such as the Hong Kong ID card,[3] and for health care or education. Such return mobilities have begun to materialize, particularly in the shape of children who are commuting across the Hong Kong–China border every day in order to attend school (Yuen 2011). As Li's (2013) research shows, even those families who do not send their HK-born children to the city for early education still anticipate that they may later go there for secondary or post-secondary study.

This case features a complex interweaving of tourism, migration and everyday travel at multiple scales. The phenomenon of 'birth tourism' was an unintended consequence of a legal judgment and a tourism policy in Hong Kong, and subsequent attempts to stop it have focused on restricting practices within both the health and immigration sectors (Cheng 2007; Deng 2012). Thinking more specifically about the experiences of families, it is clear that birth tourism sets up a complex series of mobilities. Yet these do not fit neatly within understandings of chain migration whereby family members follow each other on migration journeys (MacDonald and MacDonald 1964). Instead, a mother's 'tourism' leads to a child's 'migration'. Thereafter, family travel can become a confusing juxtaposition of both 'tourism' and 'return visits' for the citizen-child. In some cases, children are not quite 'student migrants' but more like 'educational return migrants', with 'migrations' every day, week or school term. The richness of this case is apparent, and yet studying it within a framework of 'tourism' or 'migration' would put selective emphasis upon what is a broader set of mobilities with diverse structural and personal implications.

The second example draws upon my work examining the role of materialities within the lives of return migrants (Hui 2015). While several excellent studies have discussed the importance of objects to the constitution of migrant homes and journeys (Burrell 2008, 2011; Nowicka 2007; Tolia-Kelly 2004; Walsh 2006), my project broadened the temporal scope to investigate how mobile things and people are intertwined over the course of return migration. In addition to two semi-structured interviews addressing objects that were taken and used during different migration episodes, Hong Kong-based participants completed 'object experiment books' that were inspired by the diverse creative exercises in 'probes' (Boehner, Gaver, and Boucher 2012). These books asked participants to 'follow' one object that was important in their everyday lives, discussing its mobilities past and present in both documentary and creative ways.

Methodologically, these exercises could be seen to take a significant risk. Since participants had diverse migration experiences – with some first leaving Hong Kong in the 1970s and others in the 2000s – as well as different numbers of return cycles, the time since their most recent migration varied widely. To what extent would contemporary object mobilities be relevant for an understanding of their migration? Or in what times and contexts would the mobilities of objects shed light on migration per se as opposed to other dynamics of social life? Collecting and packing things to be shipped to another country clearly links object mobilities with migration. Experiences of living abroad are also materialized when return migrants like Daniel and Jason redecorated their kitchens using appliances that were standard in previous homes but unusual in their current city. But at other times, objects have more ambiguous roles. An iPhone is a means of connecting with a daughter who has re-migrated to a country of previous residence for education, but it can also be a mobile radio that is no different in the hands of a return migrant than any other person. At those moments, it seems inconsequential that the person you are speaking to is a return migrant. By pursing an interest in both mobile people and mobile objects, this project took the risk that some of its insights might not be about migration at all. Objects are not migrants or tourists or non-migrants, though they may be moved between countries, taken on vacation and used in unremarkable parts of everyday life. Recognizing this prompts considerations of how people are also articulating between different roles, communities and institutions of relevance.

As these cases illustrate, moving beyond migrant exceptionalism is important because in the context of people's whole lives, migration is not the only frame that helps to make sense of change, adaptation, social roles, power or inequality. Neither are the social systems and diverse mobilities that affect these processes solely oriented towards 'migration'. Tourism, health and education policies can have complex implications for the regulation of migration. Expanding attention to recognize both migration and dynamics that are entangled, but beyond the realm of what is labelled migration, is therefore the important impetus for recognizing and countering migrant exceptionalism.

Some indications exist that support for moving beyond migrant exceptionalism is already building. Castles (2010, 1582–1583) has suggested the potential of re-situating theories of migration within studies of social transformation: 'A key aspect of this conceptual framework would be that it should not restrict itself to migration, but rather seek to link analysis of migratory processes to broader social theory and through this to the analysis of societal change in general'. Rogers (2005) is similarly concerned about the limits of 'migration', and points to how migrants are not the only groups involved in transnational interactions. He therefore suggests the possibility of 'folding transnationalism into a generalised study of mobilities and immobilities, in line with John Urry's "sociology beyond societies" (Urry 2000)' (Rogers 2005, 404). Any movement into such interdisciplinary spaces, however, will require careful attention to existing ontological and political differences. Favell (2007, 271–272), for example, echoes the above calls to re-situate migration studies, but suggests that migration should only become 'a subset of (spatial) mobility studies' if it can retain a focus 'on real people moving in real space – not virtual and nonhuman forms of mobility'. For Favell, the sequencing of agents is not negotiable, and people must remain ontologically primary. This is not necessarily an obstacle to interactions with mobilities research, as not all work within this field takes up non-human or virtual mobilities. Yet, it may also prompt new debates about the sequencing of concepts and concerns to, on the one hand, counter the effects of migrant exceptionalism and, on the other hand, maintain particular hierarchies of ontological status.

Regardless, there is space for further interdisciplinary engagements to grow out of attempts to counter the limitations of migrant exceptionalism. While it is possible that a lack of synchronicity or contested sequences of concepts will yet again become obstacles to further dialogue, it is also possible that this is the right time for collaborations that extend the already impressive work emerging between these fields. Engaging collaboratively around cases of sometimes-migrants could be incredibly fruitful. Whether it sparks such contributions or not, resisting migrant exceptionalism is important in order to attend to aspects of social life that have been permissible not to explore.

Conclusion

While interdisciplinary fields provide fruitful opportunities for invention and exploration, this paper has shown that their histories include much more. At the same time that engaging with shared problems or concepts can generate new contributions, it also leads to silences, struggles and attempts to limit the momentum of particular trajectories of development. Imaginings of interdisciplinary fields as open spaces for boundary crossing are therefore incomplete. Not only do these fields have various boundaries that help to define them, but they also involve attempts to define what it is permissible not to know. The possibility of interdisciplinary interactions can spark both collaboration and attempts to rule out un(inter)disciplinary concepts. A discussion of such instances has been facilitated through close attention to the circulation of concepts that are ignored, contested or inherently limited.

While one potential future trajectory is for these interdisciplinary fields to become increasingly 'disciplinary' in the management of acceptable knowledge and concepts, the paper highlighted how there is also considerable space for boundaries to change and for limitations to be challenged. Boundaries emerge within long-standing interdisciplinary fields, but they also become contested through the circulation of new concepts that highlight long-term consequences of latent assumptions. Introducing the concept of migrant exceptionalism highlights the need to reflect upon and contest assumptions about the atemporal relevance and extraordinary position of 'migrants'. Though assuming the ongoing relevance of migrants is easily done, questioning the limits of this concept is an important part of recognizing the distinction between academic practice and social life. Research traditions may perpetuate attention to certain concepts, but this does not mean that they remain the most useful ones for understanding the social world. Acknowledging migrant exceptionalism is therefore important for situating the limited relevance of 'migrants' and creating space to consider sometimes-migrants and the multiple systems their mobilities are shaped by.

This paper has also made important contributions to understandings of how temporalities are constitutive, and not only characteristic, of interdisciplinary exchanges. While cursory characterizations of interdisciplinary interactions are rhetorically convenient, they can imply that shared concerns, common concepts or ongoing dialogues are atemporal and inevitable, rather than negotiated, emergent and contingent. This discussion therefore points to the need for temporally situated and temporally sensitive accounts of interdisciplinary fields: accounts which are sensitive to a/synchronous flows of and engagement with ideas, to what is at stake in the ordering of concepts and to the accumulated limitations of even dominant frames.

Finally, this paper has added to characterizations of the intertwined histories of migration studies and mobilities research. While there has already been a significant body of collaboration between researchers in these fields, boundaries and obstacles have at times limited engagement. The discussion made clear that shared ideas are not always enough to prompt mutually beneficial dialogues. Finding new questions or problems that engage researchers from both fields will therefore be crucial for future interdisciplinary engagement between them. This paper has suggested that shared attempts to counter migrant exceptionalism and study sometimes-migrants within interlinked social systems could be one such project, which could contribute new interdisciplinary understandings of our complex, changing social worlds.

Acknowledgements

The author would like to thank James Faulconbridge, Jen Southern, Kathy Burrell and the reviewers for helpful comments on previous versions of this paper. An early draft was presented at the Global Mobility Futures conference at Lancaster University, 6 September 2013.

Notes

1. Many thanks to Kathy Burrell for bringing this point to my attention.
2. While these figures include the children of Hong Kong residents with short-term work visas, a substantial proportion of the growth is from visitors with temporary tourist visas.
3. Children over the age of 11 can apply for this card, which has a chip including biometric information and allows access to public services as well as fast-track passage through immigration checkpoints.

References

Abbott, A. 2001. *Chaos of Disciplines*. Chicago, IL: University of Chicago Press.

Adey, P., D. Bissell, K. Hannam, P. Merriman, and M. Sheller, eds. 2014. *The Routledge Handbook of Mobilities*. London: Routledge.

Ahmed, S., C. Castañeda, A.-M. Fortier, and M. Sheller, eds. 2003. *Uprootings/Regroundings: Questions of Home and Migration*. Oxford: Berg.

Bailey, A. J. 2001. "Turning Transnational: Notes on the Theorisation of International Migration." *International Journal of Population Geography* 7: 413–428.

Bailey, A. J., R. A. Wright, A. Mountz, and I. M. Miyares. 2002. "(Re)producing Salvadoran Transnational Geographies." *Annals of the Association of American Geographers* 92 (1): 125–144.

Barry, A., G. Born, and G. Weszkalnys. 2008. "Logics of Interdisciplinarity." *Economy and Society* 37 (1): 20–49.

Basch, L., N. Glick Schiller, and C. Szanton Blanc. 1994. *Nations Unbound: Transnational Projects, Postcolonial Predicaments, and Deterritorialized Nation-states*. Amsterdam: Gordon and Breach.

Basten, S., and G. Verropoulou. 2013. "'Maternity Migration' and the Increased Sex Ration at Birth in Hong Kong SAR." *Population Studies* 67 (3): 323–334.

Basu, P., and S. Coleman. 2008. "Introduction: Migrant Worlds, Material Cultures." *Mobilities* 3 (3): 313–330.

Blunt, A. 2007. "Cultural Geographies of Migration: Mobility, Transnationality and Diaspora." *Progress in Human Geography* 31 (5): 684–694.

Boehner, K., W. Gaver, and A. Boucher. 2012. "Probes." In *Inventive Methods: The Happening of the Social*, edited by C. Lury and N. Wakeford, 185–201. London: Routledge.

Boix Mansilla, V., D. Dillon, and K. Middlebrooks. 2002. *Building Bridges Across Disciplines: Organizational and Individual Qualities of Exemplary Interdisciplinary Work, 73*. Cambridge, MA: Good Work Project; Project Zero: Harvard Graduate School of Education.

Bourdieu, P. 1984. *Distinction: A Social Critique of the Judgement of Taste*. Translated by R. Nice. London: Routledge.

Bourdieu, P. 1990. *The Logic of Practice*. Translated by R. Nice. Stanford, CA: Stanford University Press.

Burrell, K. 2008. "Materialising the Border: Spaces of Mobility and Material Culture in Migration from Post-Socialist Poland." *Mobilities* 3 (3): 353–373.
Burrell, K. 2011. "Going Steerage on Ryanair: Cultures of Migrant Air Travel between Poland and the UK." *Journal of Transport Geography* 19: 1023–1030.
Büscher, M., and J. Urry. 2009. "Mobile Methods and the Empirical." *European Journal of Social Theory* 12 (1): 99–116.
Castles, S. 2007. "Twenty-first-century Migration as a Challenge to Sociology." *Journal of Ethnic & Migration Studies* 33 (3): 351–371.
Castles, S. 2010. "Understanding Global Migration: A Social Transformation Perspective." *Journal of Ethnic & Migration Studies* 36 (10): 1565–1586.
Cheng, M. H. 2007. "Hong Kong Attempts to Reduce Influx of Pregnant Chinese." *The Lancet* 369 (9566): 981–982.
Collins, F. L. 2009. "Transnationalism Unbound: Detailing New Subjects, Registers and Spatialities of Cross-border Lives." *Geography Compass* 3 (1): 434–458.
Collins, F. L. 2011. "Transnational Mobilities and Urban Spatialities: Notes from the Asia-Pacific." *Progress in Human Geography* 36 (3): 316–335.
Conradson, D., and A. Latham. 2005. "Transnational Urbanism: Attending to Everyday Practices and Mobilities." *Journal of Ethnic and Migration Studies* 31 (2): 227–233.
Conradson, D., and D. Mckay. 2007. "Translocal Subjectivities: Mobility, Connection, Emotion." *Mobilities* 2 (2): 167–174.
Crang, M., and J. Zhang. 2012. "Transient Dwelling: Trains as Places of Identification for the Floating Population of China." *Social & Cultural Geography* 13 (8): 895–914.
Cresswell, T. 2011. "Mobilities I: Catching Up." *Progress in Human Geography* 35 (4): 550–558.
Dant, T. 2004. "The Driver-car." *Theory, Culture and Society* 21 (4/5): 61–79.
Deng, A. 2012. "Loopholes in Non-resident Expectant Mother Barrier." *China Daily*. Accessed September 3, 2014. http://www.chinadaily.com.cn/hkedition/2012-03/03/content_14745471.htm%3E
Dogan, M., and R. Pahre. 1990. *Creative Marginality*. Boulder, CO: Westview Press.
Faist, T. 2013. "The Mobility Turn: A New Paradigm for the Social Sciences?" *Ethnic and Racial Studies* 36 (11): 1637–1646.
Favell, A. 2001. "Migration, Mobility and Globaloney: Metaphors and Rhetoric in the Sociology of Globalization." *Global Networks* 1 (4): 389–398.
Favell, A. 2007. "Rebooting Migration Theory: Interdisciplinarity, Globality, and Postdisciplinarity in Migration Studies." In *Migration Theory: Talking Across Disciplines*, 2nd ed., edited by C. B. Brettell and J. Hollifield, 259–278. London: Routledge.
Fortier, A.-M. 2014. "Migration Studies." In *The Routledge Handbook of Mobilities*, edited by P. Adey, et al., 64–73. London: Routledge.
Fortier, A.-M., and G. Lewis. 2006. "Migrant Horizons." *Mobilities* 1 (3): 307–311.
Franquesa, J. 2011. "We've Lost our Bearings: Place, Tourism, and the Limits of the Mobilities Turn." *Antipode* 43 (4): 1012–1033.
Garrow, D., and E. Shove. 2007. "Artefacts between Disciplines. The Toothbrush and the Axe." *Archaeological Dialogues* 14 (2): 117–131.
Gill, N., J. Caletrio, and V. Mason. 2011. "Introduction: Mobilities and Forced Migration." *Mobilities* 6 (3): 301–316.
Glick Schiller, N., and N. B. Salazar. 2013. "Regimes of Mobility across the Globe." *Journal of Ethnic & Migration Studies* 39 (2): 183–200.
Halfacree, K. 2012. "Heterolocal Identities?" *Counter-urbanisation, Second Homes, and Rural Consumption in the era of Mobilities, Population, Space and Place* 18: 209–224.
Hannam, K., M. Sheller, and J. Urry. 2006. "Mobilities, Immobilities and Moorings." *Mobilities* 1 (1): 1–22.
Holland, D. 2014. *Integrating Knowledge through Interdisciplinary Research: Problems of Theory And Practice*. Abingdon: Routledge.
Hui, A. 2012. "Things in Motion, things in Practices: How Mobile Practice Networks Facilitate the Travel and use of Leisure Objects." *Journal of Consumer Culture* 12 (2): 195–215.
Hui, A. 2015. "Networks of Home, Travel and use during Hong Kong Return Migration: Thinking Topologically about the Spaces of Human–Material Practices." *Global Networks* 15 (4): 536–552.
Kalir, B. 2013. "Moving Subjects, Stagnant Paradigms: Can the 'Mobilities Paradigm' Transcend Methodological Nationalism?" *Journal of Ethnic & Migration Studies* 39 (2): 311–327.

King, R. 2012. "Geography and Migration Studies: Retrospect and Prospect." *Population, Space and Place* 18: 134–153.
King, R., and A. Christou. 2011. "Of Counter-diaspora and Reverse Transnationalism: Return Mobilities to and from the Ancestral Homeland." *Mobilities* 6 (4): 451–466.
Lam, J., and S. Chan. 2013. "LCQ1 Annex 1: Mainland Pregnant Women Giving Birth in Hong Kong." (Press Release) Hong Kong: Legislative Council of the Hong Kong SAR Government. http://gia.info.gov.hk/general/201302/20/P201302200471_0471_107399.pdf
Levitt, P., J. DeWind, and S. Vertovec. 2003. "International Perspectives on Transnational Migration: An Introduction." *International Migration Review* 37 (3): 565–575.
Li, X. 2013. "Connect or Disconnect? Transborder Activities and Family Networks of Mainland Chinese with Children Born in Hong Kong." Paper given at David C. Lam Institute for East-West Studies Seminar Series, Hong Kong, February 25.
Lyall, C., A. Bruce, J. Tait, and L. Meagher. 2011. *Interdisciplinary Research Journeys: Practical Strategies for Capturing Creativity*. London: Bloomsbury Academic.
MacDonald, J. S., and L. D. MacDonald. 1964. "Chain Migration Ethnic Neighborhood Formation and Social Networks." *The Milbank Memorial Fund Quarterly* 42 (1): 82–97.
Mai, N., and R. King. 2009. "Love, Sexuality and Migration: Mapping the Issue(s)." *Mobilities* 4 (3): 295–307.
Meeus, B. 2012. "How to 'Catch' Floating Populations? Research and the Fixing of Migration in Space and Time." *Ethnic and Racial Studies* 35 (10): 1775–1793.
National Academies (US), Committee on Facilitating Interdisciplinary Research, Committee on Science, Engineering and Public Policy (US), National Academy of Sciences, National Academy of Engineering, and Institute of Medicine (US). 2005. *Facilitating Interdisciplinary Research*. Washington, DC: National Academies Press.
Nowicka, M. 2007. "Mobile Locations: Constructions of Home in a Group of Mobile Transnational Professionals." *Global Networks* 7 (1): 69–86.
Ohnmacht, T., H. Maksim, and M. M. Bergman. 2009. *Mobilities and Inequality*. Farnham: Ashgate.
Portes, A. 2003. "Theoretical Convergencies and Empirical Evidence in the Study of Immigrant Transnationalism." *International Migration Review* 37 (3): 874–892.
Richardson, T. 2013. "Borders and Mobilities: Introduction to the Special Issue." *Mobilities* 8 (1): 1–6.
Rogers, A. 2005. "Observations on Transnational Urbanism: Broadening and Narrowing the Field." *Journal of Ethnic and Migration Studies* 31 (2): 403–407.
Sheller, M., and J. Urry. 2006. "The New Mobilities Paradigm." *Environment and Planning A* 38: 207–226.
Skeggs, B. 2004. *Class, Self, Culture*. London: Routledge.
Strober, M. H. 2010. *Interdisciplinary Conversations: Challenging Habits of Thought*. Stanford, CA: Stanford University Press.
Tolia-Kelly, D. 2004. "Materializing Post-colonial Geographies: Examining the Textural Landscapes of Migration in the South Asian Home." *Geoforum* 35: 675–688.
Tsoi, G. 2012. "Integration Anxiety." *HK Magazine*, February, 17, 8–14.
Urry, J. 2000. *Sociology Beyond Societies: Mobilities for the Twenty-first Century*. London: Routledge.
Urry, J. 2007. *Mobilities*. Cambridge: Polity Press.
Walsh, K. 2006. "British Expatriate Belongings: Mobile Homes and Transnational Homing." *Home Cultures* 3 (2): 123–144.
Walsh, N., and H. Tucker. 2009. "Tourism 'things': The Travelling Performance of the Backpack." *Tourist Studies* 9 (3): 223–239.
Wimmer, A., and N. Glick Schiller. 2002. "Methodological Nationalism and Beyond: Nation-state Building, Migration and the Social Sciences." *Global Networks* 2 (4): 301–334.
Yuen, C. Y.-M. 2011. "Towards Inclusion of Cross-boundary Students from Mainland China in Educational Policies and Practices in Hong Kong." *Education, Citizenship and Social Justice* 6 (3): 251–264.

Mobility Infrastructures: Modern Visions, Affective Environments and the Problem of Car Parking

PETER MERRIMAN

Department of Geography and Earth Sciences, Aberystwyth University, Aberystwyth, UK

ABSTRACT *This paper builds upon recent research on mobility infrastructures to question the usefulness of the mobility/moorings binary and suggest that scholars should examine the practices of infrastructuring by which mobile subjects, affects and environments emerge. The paper outlines recent work on the affective qualities of mobility infrastructures before examining some of the diverse discourses, feelings and atmospheres which have gathered around the humble British car park over the past 50 years. It examines how the affective and experiential qualities of car parks and parking became commodified by car park designers, urban redevelopers and landscape architects, as new techniques and technologies were adopted to prevent wasteful 'space-searching' and design secure, safe, pleasant and aesthetically pleasing spaces. Finally, the paper examines some of the discourses and regulative practices which have gathered around the direct and indirect environmental impacts of parking provision and parking policies, from their influence on local ecology and hydrology, to the use of parking regulations and charges to try and reshape people's mobility habits and reduce their environmental footprints.*

Introduction

> … Stillness is so often anticipated, more or less, as an aberration and thus a problem to be dealt with. A moment of emptiness or missed productivity, producing a hobbled subjectivity without active agency. In an epoch that privileges the mobilization of mobility, still has been stilled; turned into a stop that is just waiting to go again. Waiting to be re-moved. (Bissell and Fuller 2011, 3)

Modern modes of mobility have often been associated with qualities of freedom, independence, democracy, speed and progress (Sachs 1992; Wollen and Kerr 2002;

Thacker 2003; Hagman 2006; Paterson 2007; Tomlinson 2007; Packer 2008; Seiler 2008; Duffy 2009; Merriman 2009; Rosa and Scheuerman 2009). Indeed, speed and acceleration are frequently taken as leitmotifs of modern societies, and many scholars have been drawn towards the *fast-moving* vehicle and subject, rather than the still or slow-moving vehicle, seemingly static infrastructures or mobility environments. Cutting against this obsession with movement, a broad range of other scholars have attempted to examine the role of mobility infrastructures and moorings, and qualities of slowness, materiality, durability and stillness, in both underpinning and interweaving contemporary mobilities and constituting modern societies and economies (Honoré 2004; Bissell 2007; Urry 2007; Bissell and Fuller 2011). Infrastructures enable both movement and stillness, but infrastructures are themselves dynamic, contingent, in process and in movement (Adey 2006; Merriman 2012). In this paper, I want to question both the ontological status of infrastructure and the binary of mobility/moorings, urging scholars to shift their attention from accounts of fixed and bounded 'infrastructures', to develop more detailed understandings of the relational processes of 'infrastructuring' by which infrastructured spaces, subjects and practices emerge.

In section one of this paper, I outline why mobility scholars should undertake such a post-structuralist, post-humanist and processual move; outlining recent work on the processes and affects of infrastructuring, including writings on vibrant materialities, mobile affects and mobile methods. In section two, I trace the birth of the modern and modernist car park in Britain, revealing how new car parks were seen as a solution to an emerging 'parking problem'; forming an important part of the post-war redevelopment of many towns and cities. I examine how conservationists began to criticise the effects that large car parks had on the townscape of cities in the 1970s, before tracing how Britain's stark modernist multi-storey car parks emerged as settings for a range of artistic and fictional events – from murders in films, to staged performances. In section three, I examine how the affective and experiential qualities of parking infrastructures were commodified by car park designers, urban redevelopers and landscape architects, as new techniques and technologies were increasingly adopted as ways to prevent wasteful 'space-searching', re-brand these spaces, and design secure, safe, pleasant and aesthetically pleasing spaces. Finally, I examine some of the discourses which have gathered around the direct and indirect environmental impacts of parking infrastructures and parking policies, from their influence on local ecology and hydrology, to the increasing use of parking regulations in trying to reshape people's mobility habits and reduce their environmental footprints.

Mobility Infrastructures or Mobile Infrastructuring?

Mobility scholars have written extensively on mobility infrastructures, as can be seen in successive issues of *Mobilities* over the past decade. In the editorial to volume 1, issue 1, Hannam, Sheller, and Urry (2006) drew a distinction between mobilities and the Lefebvrian concept of 'moorings' (Lefebvre 1991), conceived as 'the necessary spatial, infrastructural and institutional moorings that configure and enable mobilities' (Urry 2003; Hannam, Sheller, and Urry 2006, 3). Mobilities, it was argued, rely upon all manner of fixities, and an array of scholarly work has focused on such seemingly static mobility infrastructures as roads, railways, airports, ports and urban infrastructures, as well as more mobile infrastructural assemblages such as shipping containers and various types of vehicle (e.g. Graham and Marvin 2001; Merriman 2007; Lipset and Handler 2014). Mobility infrastructures are frequently discussed in terms of the spaces, sociabilities, practices, qualities and politics associated with them, but in this paper I want to discuss the implications of a rather different approach to infrastructural matter, materiality and practices, arguing that realist approaches that naturalise the concrete reality of infrastructures may lead us to overlook the affective and atmospheric qualities of such spaces and oversimplify the ways in which qualities and practices resonate with and cut across people and things, spaces and subjects, matter and thought, and what some have termed mobilities and moorings.

A broad range of scholars – including transport historians, cultural geographers, sociologists, anthropologists, architectural theorists, communications scholars, urban theorists and ANT/STS scholars – have held a fairly long-standing interest in the physical infrastructures and materialities of mobility and transport networks. While some scholars have approached infrastructures as material technologies which structurally determine socio-spatial relations and practices, contemporary thinkers more frequently emphasise the non-linear, networked, relational and distributed qualities of infrastructures and mobility spaces and practices. In the case of geographers and a few anthropologists, there has been a particular focus not only on how transport routes striate the landscape, but also on how mobility, communications and transport infrastructures necessitate a rethinking of Euclidean and topographic senses of space, as they generate uneven topologies of connection and disconnection, mobility and immobility (Harvey 1989; Massey 1991; Harvey 2012; Merriman 2012). For example, cultural geographers, anthropologists and other mobility scholars have examined how limited-access routes such as motorways, railway lines and air routes generate 'corridoring' or 'tunnelling' effects in landscapes (Andreu 1998; Graham and Marvin 2001; Bishop 2002; Merriman 2011; Dalakoglou and Harvey 2012). In discussing her relational and dynamic approach to space as 'a configuration … of a multiplicity of trajectories', Doreen Massey (2000, 225) has similarly emphasised how a myriad of trajectories of people and things comprises and refigures our ever-changing dynamic worlds, while an increasing number of other scholars are drawing upon relational and post-humanist approaches to materiality, nature, landscape and technologies in an attempt to expose the emergence of socio-material assemblages, mixtures, configurations or hybrids which problematise any idea of stable, fixed and neatly bounded infrastructures (cf. Latour 1993; De Landa 2006).

In many ways, these new approaches to infrastructure present very different arguments to those made by leading mobility scholars. In his book *Mobilities*, John Urry has stressed the 'complex character' and 'powerful spatial fixity' of systems of mobility which 'structure mobility experiences' (Urry 2007, 54, 52, 54), and in both *Mobilities* and his earlier book *Global Complexity*, he refers to the 'dialectic of moorings *and* mobilities' (Urry 2003, 138). Movement and fixity appear as common sense, perceptible, real-world 'states' or 'conditions', demonstrating how the 'new mobilities paradigm … advocates for a realist relational ontology for contemporary social science' (Sheller 2014, 790). Cresswell (2014a, 719) also appears to resort to a realist ontology when arguing that 'immobilities (and indeed time-spaces) … cannot be wished away with a theoretical wand', but others have questioned the construction of simple binaries and of allusions to 'realism'. Writing in the journal *Mobilities* back in 2006, Peter Adey suggested 'that there is never any absolute immobility, but only mobilities which we mistake for immobility, what could be called relative immobilities', and as such 'moorings are indeed mobile too' (Adey 2006, 83, 86). While I would echo Adey's (2006) comments, I would go further, suggesting that the very binary or dialectic of mobility/moorings is problematic, and that realist constructions of mobility – in focusing on particular manifestations and understandings of human embodied movement, and 'common sense' observations of largely visible 'molar' movements – overlook an array of less perceptible, obvious, irreducible or impending 'molecular mobilities' (Merriman 2012, 3–7), as well as vibrant materialities and mobilities which might be said to saturate and constitute our worlds (Bennett 2010; Bissell 2010a). There is a danger that 'common-sense' realist imaginations (even if driven by important political imperatives) merely reinforce Western epistemologies, ontologies and binaries, as well as overlooking the multiplicity of mobilities, temporalities and spatialities emerging from the unfolding of events. In contrast, important work has emerged both within and beyond mobility studies which seeks to understand the affective resonances, atmospheres, vibrations and enchanting qualities of mobility infrastructures and practices.

Firstly, there has been an emergent literature on situations, experiences and atmospheres of relative stillness, quiet and apparent inactivity, in which stillness is not seen to be the opposite of movement, but 'might [be seen to] emerge through other configurations of matter which are not necessarily reducible to the dialectic of mobility and immobility' (Bissell and Fuller 2011, 6). Here, there is an attempt to advance a more relational argument around questions of movement, stillness and rhythm, which does not rely upon a simple binary of mobility and fixity. Movement is not simply registered or measured as physical displacement, speed or acceleration. Rather, movements may have very different qualities, temporalities, rhythms, materialities and affective resonances, which cut across and resonate through things we know and separately identify as infrastructures, environments, vehicles and bodies. Movements vibrate, resonate and circulate, but these movements are not without politics or political manipulation, and neither are they nomadic, smooth or chaotic. Mobility systems are characterised by degrees of friction and turbulence (Cresswell and Martin 2012; Cresswell 2014b). The elemental materialities of mobility infrastructures are often fairly predictable, while it is not just mobile subjects (but also powerful political and economic actors) who recognise how both mobility infrastructures and practices get caught up in affective relations and atmospheres. Mundane infrastructures such as roads are 'charged composition[s]'of 'bodies, rhythms, orientations, and atmospheres' (Stewart 2014, 552, 549). Roads vibrate with affective atmospheres (cf. Bissell 2010a, 2010b), and they have the 'capacity to enchant' (Harvey and Knox 2012, 521).

Secondly, mobile methods can provide one important means for examining how mobility infrastructures, environments and subjects become entangled and entwined in complex ways, as complex ontological assemblages emerge in the unfolding of *particular* movements, and distinctive rhythms, auras, atmospheres, affects and tensions take hold in bodies-in-movement-in-vehicles in particular infrastructural environments. In a range of studies utilising and advocating techniques such as video ethnography, go-along interviews and participatory and practice-based research, academics have shown how particular practices, experiences, decisions, interactions, feelings and affects resonate between mobile subjects, vehicles, infrastructures and landscapes (Bissell 2010a, 2010b; Fincham, McGuinness, and Murray 2010; Büscher, Urry, and Witchger 2011). The immersive infrastructures and environments through and with which people move are central to experiences of railway travel (Watts 2008; Bissell 2010b), cycling (Spinney 2006), or driving and passengering (Laurier et al. 2008), and they are not simply static contextual settings for action, but are actively produced and shaped through embodied movements, interactions and techniques of inhabitation (Massey 2005; Merriman 2012). Of course, mobile methods are not the only way to understand these dynamic, embodied, mobile practices (Merriman 2014), but they can provide important illuminations of the messy unfolding of mobilities, in which dynamic practices of moving and stilling entail complex processes of infrastructuring and environing, as mobile worlds emerge from these processual unfoldings – an insight one can trace back through traditions of phenomenology, process philosophy and non-representational theory (Merriman 2012).

What I am suggesting is that the affective power of infrastructures extends far beyond their immediate physical bounds, and that infrastructures are in constant process – both socially and materially – requiring physical maintenance, gathering meanings and generating atmospheres (Graham and Thrift 2007; Harvey 2012; Harvey and Knox 2012). Our focus should be on different practices of 'infrastructuring' rather than simple geographies of infrastructures, and in this paper I want to highlight the multiple temporalities, intermittent practising, and sustained periods of stillness and quiet associated with the practices and spaces of parking in the UK. Indeed, while parking has long been a concern of policy-makers and transport policy, planning, land use, architecture and economics academics (Baker and Funaro 1958; McCluskey 1987; Jakle and Sculle 2004; Henley 2007; Shoup 2011; Ben-Joseph 2012; Marusek 2012), these practices and spaces have attracted very little attention from mobility scholars (although see Hagman 2006; Henderson 2009; Larsen 2015). What many studies emphasise is the very different parking policies, provision and land use pressures which have emerged in different societies, most notably with the vast scale of parking lots in some US cities, and the future parking challenges posed by the expansion of car ownership in countries such as China (Ben-Joseph 2012). In this paper, however, my focus is on the different ways in which parking spaces become caught up in cultural, economic, political and environmental practices and relations. Parking practices draw upon a range of way-finding techniques and technologies, as well as knowledge and judgements about where might be the cheapest, safest and most convenient places to park. We may avoid particular car parks because they are too expensive, congested, have narrow spaces, or because they feel unsafe or unsecure. We may shop in out-of-town shopping centres because they are accessible and parking is straightforward and free. We may rent or buy a house because it has a garage, driveway or unrestricted parking. We may be persuaded not to use a private car due to the absence of parking infrastructure or its cost. We may get angry about parking regulations or the parking habits of others (Marusek 2012). What all of these different associations (and more) highlight is how parking infrastructures generate complex affects, experiences and atmospheres, 'infrastructuring' practices, subjects, atmospheres and environments in diverse ways. What's more, like most public places, car parks are not simply used and consumed in ways intended by their designers and operators, being multi-functional spaces with distinctive temporalities. This may include their use by 'boy racers' (Lumsden 2013), skateboarders (Borden 2001), car-boot sale organisers (Gregson and Crewe 1994) or by 'doggers' seeking to watch, be watched or partake in public sex, often in cars, usually at night (Bell 2006). But why did these infrastructuring practices emerge in the first place?

The Birth of Parking? Modernist Dreams and Urban Nightmares

Early motorists in the 1890s and early 1900s had to think carefully about where to stop, as early motor cars were generally open-topped and rarely featured locks or ignition keys (O'Connell 1998; Merriman 2012). This, coupled with the laws on obstructing highways, meant that many urban motorists in Britain chose to 'stable' their cars in specialist garages, rather than to park them by the road side (Morrison and Minnis 2012, 172). As covered cars became more popular in the 1910s and 1920s – and as security devices were improved as a result of the increasing occurrence of joy-riding (O'Connell 1998) – so motorists parked their cars on public streets in increasing numbers for longer periods, but alongside this increased presence came criticisms of the nuisance, obstructions and visual chaos caused by parked vehicles. By the 1940s, the streets of London were seen to be unacceptably congested with parked cars, and a series of experimental surveys were undertaken by the Ministry of Transport and Civil Aviation and Road Research Laboratory to develop an understanding of people's parking habits and identify the causes and solutions of the 'parking problem' (Ministry of Transport and Civil Aviation 1956, 1958; Buchanan 1958). At the heart of these studies were fairly basic ideas of supply and demand economics, and somewhat simplistic assumptions about people's parking habits. The key challenges foreseen by government scientists and civil servants were how to provide infrastructures for increasing numbers of motorists, how to free up on-street parking space for short-term visits, and how to incentivise local authorities and private companies to construct off-street car parks that would increase capacity (Ministry of Transport 1963). At the heart of their approach was an assumption that infrastructures would not only structure environments, but that they could influence people's decisions and habits. One key solution, installed from 1958, was the mechanical, coin-operated parking meter, which Minister of Transport Ernest Marples believed imposed the 'discipline and control needed to keep traffic moving' (Marples 1963, 4).

While the government sought to enable and speed up parking through various policies, regulations and new infrastructures, some urban thinkers questioned the provision of more roads and parking spaces. Writing in *The Architectural Review* in 1958, the editor and renowned critic J. M. Richards suggested that an 'alternative policy' would be to discourage the use of motor cars and engender 'public enthusiasm for public transport' (Richards 1958, 299), while the US urbanist Lewis Mumford expressed similar concerns about London in 1961:

> … the motorcar, floating in on a wave of prosperity, has clogged the highways and byways of the city. This congestion has brought with it the conventional train of expert pseudo-remedies, from parking meters to the chipping away of space in squares and parks, above all, in Hyde Park. But all that – besides reducing the acreage of precious recreational areas, transforming building lots into parking lots, and inviting still more traffic – simply ensures that movement will eventually come to a standstill unless the combined remedies bring about a total evacuation of the central core of the city. (Mumford 1964, 118)

The government, local authorities and the ever-increasing motoring public saw the motor car and new roads as convenient solutions to Britain's congestion problems, but the proliferation of street furniture, signage and new car parks were also seen to have negative effects on the urban townscape and rural landscape (Nairn 1955; Cullen 1961; Middleton 1962; Merriman 2007). With the increasing focus of local authorities on reconstructing Britain's town and city centres, multi-storey car parks started to emerge as a distinctive, new, modernist architectural form in the urban landscape, alongside the construction of modernist shopping centres, schools and office blocks in the 1960s (Morrison and Minnis 2012). Encouraged by central government, local authorities and private developers constructed hundreds of surface and multi-storey car parks as part of their redevelopment plans for many British town and city centres, but by the early 1970s these transformations of the urban environment were attracting widespread criticism from an emerging conservation movement. To take just one criticism of a single town, described by Colin Amery and Dan Cruickshank in *The Rape of Britain*:

> The small-scale cobbled streets of central Gravesend now survive only as boundaries between fields of tarmac. All has been levelled to provide ample acres for the visiting motorist to park his car. (Amery and Cruickshank 1975, 86)

In their quest to maximise parking and cater for shoppers and tourists, many local authorities were criticised for their ambitious parking schemes, with Canterbury City Council receiving objections from conservationists and architects after they proposed a seven-storey car park as part of a new development, just 400 yards from the cathedral (*The Times* 1968a, 1968b). Stark, brutalist concrete multi-storey car parks quickly moved from being seen as exciting futurist experiments to being received as rather intrusive, ugly, dystopian impositions, which produced negative affects, feelings and atmospheres. Science fiction writers such as JG Ballard populated their dystopian worlds with such spaces, with multi-storey car parks becoming a favoured site for illicit sexual encounters (Ballard 1973).

One of Britain's most stark and iconic multi-storeys was Trinity Square car park in Gateshead, designed in 1967 by Rodney Gordon for Owen Luder Partnership. This monumental car park became famous for being the site of a murderous encounter in the 1971 film *Get Carter* starring Michael Caine, as well as attracting more recent attention from artists and film-makers, up until its demolition in 2010. In 1983, Trinity Square was the site of Charlie Hooker's performance installation *MAINBEAM (A Ballet for Vehicles)* for The Basement Group wherein cars and pedestrians were choreographed to move 'toward[s] and away from the audience and gradually, menacingly, hemming it in' (Hooker 2007, 48). The location, infrastructural affects, and atmosphere of this unpopular and run-down car park were central to the performance, for as Hooker later reflected:

> In MAINBEAM, the idea was to take an audience to an extremely threatening and austere location, create a performance and push this to a level where as a viewer each person began to question at what point performance ended and reality began. (Hooker 2007, 48)

Modern brutalist multi-storey car parks vibrate and resonate with a complex array of affects and atmospheres – from oppressive, dystopic affects, to nostalgic feelings – and these atmospheres are in turn re-engineered by artists, film-makers and performance artists. There is, of course, a rich tradition of performance and visual artists using vehicles in their work (Pearson and Shanks 2013), but the aesthetics, acoustics and spatialities of the humble 'car park' has captured the imaginations and served as a stage for a range of artistic endeavours, from photographer Martin Parr's photographic study of parking spaces (Parr 2007), to the classical concerts staged by London group 'Multi-Story' in Peckham Rye multi-storey car park, every summer since 2011 (Whitley and Stark 2014).

Affective Environments and Economies

Parking infrastructures are bound up with complex affective economies, decision-making processes and aesthetic judgements, whether by designers, operators or drivers. Parking spaces are frequently 'valued' by drivers when they are convenient, accessible, secure, well-designed and economical, but they are an unusual and subtly variable global commodity with few tangible characteristics beyond the available space, land or territory. It is these qualities which inspired the British photographer and cataloguer of everyday life Martin Parr to engage in a quest to photograph 'the last parking space' at numerous sites around the world between 2002 and 2007. At a conceptual level, Parr was fascinated by these highly sought after yet elusive globalised commodities: 'the one thing that we're all looking for in life is somewhere to park the car' (Parr 2003). Parr photographed 'the last parking space' in different locations in 41 countries, ranging from Argentina and India, to Syria and Singapore, documenting them in a large format, boxed white faux-suede album resembling a luxury wedding album, 'suggesting the precious nature of the elusive parking space' (Parr 2014).

Motorists and a range of cultural commentators clearly value parking spaces, and in congested areas motorists often utilise a range of local knowledges and way-finding techniques to find a space – coupled with elements of happenstance and luck. Modern technologies such as mobile phone applications, satellite navigation systems, websites and digital roadside information screens are increasingly supplementing road signs, guides and paper maps as ways to locate parking spaces and direct motorists, cutting down the time and associated congestion and pollution of 'space-searching' (Shoup 2006, 2011). In a number of British towns and cities, occupancy sensors relay the number of available spaces in different car parks to signs on major approach roads, while in San Francisco (USA), city authorities have introduced 'performance pricing', with parking sensors and variable pricing being used to manipulate occupancy rates so they achieve the optimum of about 85% (Shoup 2011, xx–xxv). Private property owners are increasingly leasing unused parking space through websites such as Parkatmine.com, Parklet.co.uk and Parkatmyhouse.com, enabling the online marketing and selling of unused spaces at homes and offices in prime locations. Parking spaces have a long history of commodification and valuation, but the increasing demand for safe and secure parking, coupled with the rise of new technologies – whether for space finding, leasing, occupancy monitoring, surveillance or payment processing – has led to a transformation of the geographies and economies of parking in recent years.

Safety and security are key qualities which have perhaps always been valued by motorists parking their cars. Indeed, car parks have often been experienced as marginal, threatening and indeed dangerous places, particularly at night, and often by women (Valentine 1989). Good car park design is seen to be vital to ensuring the safety and security of users and their vehicles, and car park designers have long advocated the application of design principles which will increase natural as well as artificial surveillance, with Oscar Newman's pioneering research on 'defensible spaces' being seen to be easily applicable to car park design (Newman 1973; McCluskey 1987). Since 1988, the UK Association of Chief Police Officers have promoted a number of schemes around the branded concept 'Secured by design', including their 'Park Mark® Safer Parking' scheme (run by the British Parking Association), which is marketed as a tool by which operators can demonstrate to their 'customers a commitment to creating a high quality parking facility, that feels safe and has measures in place to reduce levels of crime' (BPA 2013). Safety and security are seen to be vital qualities to *design into* new car parks, designing affective environments which generate feelings and atmospheres of safety and security (cf. Adey 2008; Anderson 2014). In the 'Park Mark®' scheme, it is made clear that new car parks should be clearly bounded territories, well-lit, have good sight lines, good signage and often CCTV systems, although expert site inspections form the heart of the accreditation process, taking account of local conditions (BPA 2010). Parking infrastructures are also seen as potential security risks from bomb threats, and in 2012, the UK Home Office, along with the Centre for the Protection of National Infrastructure and National Counter Terrorism Security Office, issued detailed design guidance on how to protect 'crowded places', including advice on the design and security of car parks (Home Office 2012). Car parks must be kept secure, whether from terrorists, thieves or vandals.

Decisions surrounding parking provision have been vital to architects, planners and engineers seeking to design safe, pleasant and desirable urban and rural developments since at least the 1950s. Decisions about levels of parking provision and parking location, design and landscaping are seen to be vital to the success of schemes, with research by the UK Department of the Environment on tenant satisfaction in public housing schemes in the early 1970s revealing people's strong views on parking, with them preferring 'uncovered parking spaces within sight of the owner's dwelling combining, as it does, low rent, security and proximity', as 'cars parked on open, grouped, hard standings away from the house are often subject to vandalism' (McCluskey 1987, 34). More recently, the Department for Transport's (2007) *Manual for Streets* recommends small clusters of courtyard parking for security and good visual design, while a range of techniques of landscaping and planning have been promoted as ways to 'civilize the parking precinct' and 'soften the ambience' of areas populated with parked cars (McCluskey 1987, 16, 17). Many planners and engineers have moved away from the construction of large-scale tarmac car parks with uniform planar surfaces, advocating alternative design and landscaping strategies in an attempt to create more intimate spaces, positive atmospheres and break up the visual monotony of expansive and impervious surfaces (Ben-Joseph 2012). In the late 1980s, designers such as Jim McCluskey encouraged planners, architects and engineers to approach car parks as 'an outside room' or 'contained outside space', designing them in three dimensions and utilising the 'roofing effect' provided, for example, by trees (McCluskey 1987, 4, 15).

Since the 1990s, parking areas have been redesigned as part of broader attempts to develop streets as shared infrastructures or 'shared spaces' (Ben-Joseph 1995). The removal of clear demarcations between pedestrian and traffic areas places emphasis on 'informal social protocols and negotiation', re-introducing the challenges and unpredictable movements of mixed traffic which existed for centuries before the introduction of traffic segregation technologies in cities (Hamilton-Baillie 2008, 166). While the removal of curbs, painted lines, signage and rights of way is intended to improve the aesthetics of these spaces, promote caution and responsibility, and enact positive behaviour change amongst motorists and other road users, motorists do still need to be clearly informed about where they can park and for how long (DfT 2011a). The consequences of not doing this include confused motorists and illegal parking – which have been observed in some shared space schemes (see Cornwell 2013). Parking areas need to be coherent and 'legible' for motorists. What 'shared space' designers, car park operators and landscape designers appear to agree is that parking areas should be visually pleasant environments that generate positive affects and atmospheres, but it is not just the aesthetic aspects which concern many commentators. Parking infrastructures may also have a significant impact on the natural environment.

Parking Infrastructures and 'The Environment'

Parking infrastructures and practices have a broad range of direct and indirect impacts on the environment. The more direct environmental impacts of parking include 'degraded water quality, stormwater management problems, exacerbated heat island effects, and excessive land consumption' (EPA 2006, 11), which have been known by physical scientists, environmentalists and architects for decades. Attention has often been paid to the direct environmental impacts of *large* public and private car parks, whether visually or in terms of their impact on hydrology and ecology, but planners, environmentalists and academics have also expressed concern about the cumulative impact of private off-road parking areas on the environment. In the UK, this has included concerns about the potential impact that increasing numbers of remodelled and paved domestic gardens may be having on drainage, flooding and urban green space (Smith 2010; Warhurst et al. 2014). One technological solution which was popular amongst car park designers, architects and landscape designers during the 1970s and 1980s were alternative surfacing and 'greening' technologies, especially in low use, seasonal and rural public car parks. In contrast to impermeable tarmac and concrete surfaces, designers have experimented with a range of permeable and semi-permeable surfaces with varying appearance, durability and drainage properties, ranging from reinforced grass surfaces, to patented hybrid surface systems such as Grasscrete®, Hexpot® and Bodpave® in which grass growth and water drainage are enabled and supported with patterned concrete or plastic structures (e.g. McCluskey 1987, 170–171).

The direct environmental costs of parking provision and car park design are clearly significant, but the indirect environmental costs of parking policies have the potential to have more widespread and cumulatively catastrophic impacts on the environment and global climate. In a 2006 report entitled *Parking Spaces/Community Places*, the US Environmental Protection Agency explained how parking 'indirectly affects the environment, primarily because parking influences how and where people choose to travel' (EPA 2006, 11), while the UK Department for Transport's *Manual for Streets* emphasised that 'levels of parking provision and its location has a key influence on the form and quality of a development, and the choices people make in how they travel' (DfT 2007, 99). Parking policies can form a key component of local and national environment and climate change mitigation policies, and transport researchers have undertaken extensive research to try and assess the role of parking policies, restrictions and provision in people's mobility decisions and modal choice (e.g. Tsamboulas 2001; Kodransky and Hermann 2011). At a time when policy-makers and politicians in Britain and other developed nations are increasingly advocating and adopting 'behaviour change' techniques as a way to govern citizens and shape their habits (see Jones, Pykett, and Whitehead 2013), revised parking policies have emerged as a key technique by which local government authorities, transport planners and employers are seeking to reshape people's mobility habits, bring about changes in behaviour, and '(infra)structure' our lives in new ways. Within these discourses, behaviour tends to be approached through psychological or even neurological approaches wherein habits are 'apprehended as resources located within individual bodies that are at the mercy of disruption from the outside', whereas more distributed approaches 'carve out a much more "displaced" understanding of habit where habit does not reside within bodies but is the more creative and unpredictable process through which bodies and environments mutually evolve and take shape through the dual process of the taking on of, and annihilation of, form' (Bissell 2013, 121). In the latter approach, habits are seen to be produced and enacted through bodies-in-vehicles moving through and inhabiting particular infrastructures and environments. Habits and environments are, in effect, shaped through multiple 'infrastructuring' practices aimed at constituting and remodelling bodies and/in environments.

Policy-makers and local authorities are increasingly experimenting with different approaches which they believe may bring about changes in people's parking, mobility and environmental habits, ranging from modified parking provision and public transport infrastructures to revised parking pricing strategies. Of course, environmental parking policies are not entirely new. Park-and-ride schemes first appeared in the UK in the 1960s, although not in large numbers until the 1990s (Manns 2010), being framed as a solution to urban congestion and latterly as a way to reduce urban pollution and bring about modal shift – despite limited evidence of their effects on car journeys, and suggestions that they could lead to increased car traffic (Parkhurst 1995). More recently, policy-makers across Europe have been developing parking policies which encourage travellers to shift mode and bring about compliance with EU air quality and emissions targets (Kodransky and Hermann 2011). These include so-called environmental or emissions-based parking charges, which have emerged as one strategy adopted in certain areas of the UK (Wall 2011). Local councils such as Westminster in London and Winchester in Hampshire have introduced variable annual parking permit charges depending on the CO_2 emissions of a vehicle (which is also linked to the vehicle road tax band charged by the UK Government). In Westminster, owners of 'eco'/electric vehicles and car clubs can apply for free parking permits, while in local authorities such as Winchester they receive significant discounts (Wall 2011). Environmental parking policies such as these are intended to provide incentives for motorists to use public transport, purchase low-emissions vehicles or join car clubs. Likewise, the UK Government's 'Plug-in Vehicle Strategy' contains a range of other measures designed to promote improvements in electric vehicle parking provisions and charging infrastructures, helping to promote more eco-friendly vehicles (DfT 2011b). Workplace parking levies – introduced by Nottingham City Council, but not adopted elsewhere in the UK – are another policy designed to reduce urban congestion, channel funds into public transport schemes and encourage commuters to switch to more environmentally friendly private and public transport modes (Nottingham City Council 2011). Levies are made on employers for every employee parking space they have above 10 (these fees were £379 per parking place per year in April 2015 in Nottingham); fees which employers can pass on to employees as part of an annual parking charge. In the case of the University of Nottingham, a staff parking permit scheme was introduced to 'support the University's environmental agenda and … meet the costs of the City Council Workplace Parking Levy', with fees being scaled depending on staff salary levels as well as vehicle emissions rates (University of Nottingham 2013). The committee controlling this university permit scheme have expressed concern about the differential impact that parking fees can have on low-paid employees and part-time employees. Like some other workplace parking schemes, they have attempted to consider economic and social factors which cannot be easily accounted for or enforced in public parking schemes, where special arrangements are only really in place for motorists with mobility and accessibility issues (with parking reserved for people with physical disabilities and parents with young children).

Conclusions

> … a successful parking lot is one that integrates its site conditions and context, takes measure to mitigate its impact on the environment, and gives consideration to aesthetics as well as the driver-parker experience. Designed with conscientious intent, parking lots could actually become significant public spaces, contributing as much to their communities as great boulevards, parks, or plazas. (Ben-Joseph 2012, 136)

Early car parks were often fairly basic, functional modernist infrastructures, but contemporary car park designers and urban theorists appear to acknowledge the multiple and often complex infrastructuring processes and affects that these spaces can generate. In rethinking car park design in tandem with transport and urban policies, a broad range of experts and authorities are becoming increasingly concerned with trying to re-engineer and restructure mobility habits and environments (or habits-in-environments), infrastructuring the world in new ways. For city planners and urban designers such as Eran Ben-Joseph, we must rethink the design and function of parking infrastructures so that they are elevated above their lowly status as functional buildings – or perhaps as monotonous 'in-between' spaces or 'non-places' – which we at best tolerate or overlook, and at worst despise (cf. Henley 2007). The key is to enfold parking facilities into comprehensive architectural and landscape designs, developing coherent and meaningful shared spaces in which parking practices and spaces are slowed and scaled down, re-centred and re-humanised, rather than banished to the margins. Indeed, there is a widespread acceptance that parking practices and spaces are important to the functioning of contemporary automobilised economies and societies, with the regulation, control and commodification of parking becoming important practices for both private companies and local governments, whether for economic, political, environmental or congestion reasons. At a time when national and local governments are looking for innovative and 'acceptable' ways to control the use of high CO_2 emissions petroleum-fuelled vehicles, parking regulations appear to be a more politically acceptable way to try to limit and restructure car use and enact behaviour change than strategies such as fuel taxation. A broad range of policy regulations have provided strategies for attempting to 'tip the balance' in favour of public transport and low-emissions private vehicles, whether through transforming the material architectures and affective qualities of car parks, or attempting to transform the economic performance of these commodified spaces. What is clear is that parking 'matters', generating diverse infrastructural affects. Parking infrastructures and policies are regularly a subject of comment and observation by motorists, environmentalists, designers and transport experts, as well as local residents, cultural commentators, artists and performers, whether in complaining about local authority charging regimes, reflecting on people's search for the banal, globalised parking space (Parr 2007), or utilising car parks as settings for dark stories (Ballard 1973) or atmospheric performances (Hooker 2007; Whitley and Stark 2014). Parking both excites and infuriates, and local and national governments face major challenges when seeking to shape mobility decisions and habits through parking policies as part of much broader attempts to shape mobility habits and desires, and enact modal shifts. Parking spaces are not static and stilled infrastructures, rather they are complex environments which gather and resonate with multiple affects and atmospheres, 'infrastructuring' subjects, environments and communities in diverse ways.

References

Adey, P. 2006. "If Mobility is Everything Then It is Nothing: Towards a Relational Politics of (Im) Mobilities." *Mobilities* 1: 75–94.

Adey, P. 2008. "Airports, Mobility and the Calculative Architecture of Affective Control." *Geoforum* 39: 438–451.

Amery, C., and D. Cruickshank. 1975. *The Rape of Britain*. London: Paul Elek.

Anderson, B. 2014. *Encountering Affect*. Farnham: Ashgate.

Andreu, P. 1998. "Tunneling." In *Anyhow*, edited by C. Davidson, 58–63. Cambridge, MA: The MIT Press.

Baker, G., and B. Funaro. 1958. *Parking*. New York: Reinhold Publishing.

Ballard, J. G. 1973. *Crash*. London: Jonathan Cape.

Bell, D. 2006. "Bodies, Technologies, Spaces: On 'Dogging'." *Sexualities* 9 (4): 387–407.

Ben-Joseph, E. 1995. "Changing the Residential Street Scene: Adapting the Shared Street (Woonerf) Concept to the Suburban Environment." *Journal of the American Planning Association* 61 (4): 504–515.

Ben-Joseph, E. 2012. *Rethinking a Lot*. London: The MIT Press.

Bennett, J. 2010. *Vibrant Matter*. London: Duke University Press.

Bishop, P. 2002. "Gathering the Land: The Alice Springs to Darwin Rail Corridor." *Environment and Planning D: Society and Space* 20: 295–317.

Bissell, D. 2007. "Animating Suspension: Waiting for Mobilities." *Mobilities* 2 (2): 277–298.

Bissell, D. 2010a. "Vibrating Materialities: Mobility-Body-Technology Relations." *Area* 42: 479–486.

Bissell, D. 2010b. "Passenger Mobilities: Affective Atmospheres and the Sociality of Public Transport." *Environment and Planning D: Society and Space* 28 (2): 270–289.

Bissell, D. 2013. "Habit Displaced: The Disruption of Skilful Performance." *Geographical Research* 51: 120–129.

Bissell, D., and G. Fuller. 2011. "Stillness Unbound." In *Stillness in a Mobile World*, edited by D. Bissell and G. Fuller, 1–17. London: Routledge.

Borden, I. 2001. *Skateboarding, Space and the City*. Oxford: Berg.

BPA (British Parking Association). 2010. *New Build Car Park Guidelines for Car Park Designers, Operators and Owners*. Haywards Heath: Safer Parking Scheme.

BPA (British Parking Association). 2013. *"Park Mark® Safer Parking" Leaflet*. Haywards Heath: Safer Parking Scheme.

Buchanan, C. 1958. *Mixed Blessing*. London: Leonard Hill.

Büscher, M., J. Urry, and K. Witchger, eds. 2011. *Mobile Methods*. London: Routledge.

Cornwell, R. 2013. "Felixstowe: Nearly 900 Drivers Fined for Illegal Parking in Resort's Shared Space." *Ipswich Star*. Accessed March 23. http://www.ipswichstar.co.uk
Cresswell, T. 2014a. "Mobilities III: Moving On." *Progress in Human Geography* 38: 712–721.
Cresswell, T. 2014b. "Friction." In *The Routledge Handbook of Mobilities*, edited by P. Adey, D. Bissell, K. Hannam, P. Merriman, and M. Sheller, 107–115. London: Routledge.
Cresswell, T., and C. Martin. 2012. "On Turbulence: Entanglements of Disorder and Order on a Devon Beach." *Tijdschrift voor Economische en Sociale Geografie* 103: 516–529.
Cullen, G. 1961. *Townscape*. London: The Architectural Press.
Dalakoglou, D., and P. Harvey. 2012. "Roads and Anthropology: Ethnographic Perspectives on Space, Time and (Im)Mobility." *Mobilities* 7 (4): 459–465.
De Landa, M. 2006. *A New Philosophy of Society*. London: Continuum.
DfT (Department for Transport). 2007. *Manual for Streets*. London: Thomas Telford.
DfT (Department for Transport). 2011a. *Shared Space (Local Transport Note 1/11)*. London: Department for Transport.
DfT (Department for Transport). 2011b. *Making the Connection: The Plug-in Vehicle Infrastructure Strategy*. London: Department for Transport.
Duffy, E. 2009. *The Speed Handbook*. London: Duke University Press.
EPA (United States Environmental Protection Agency). 2006. *Parking Spaces/Community Places*. Washington, DC: Environmental Protection Agency.
Fincham, B., M. McGuinness, and L. Murray, eds. 2010. *Mobile Methodologies*. Basingstoke: Palgrave Macmillan.
Graham, S., and S. Marvin. 2001. *Splintering Urbanism*. London: Routledge.
Graham, S., and N. Thrift. 2007. "Out of Order: Understanding Repair and Maintenance." *Theory, Culture and Society* 24 (3): 1–25.
Gregson, N., and L. Crewe. 1994. "Beyond the High Street and the Mall: Car Boot Fairs and the New Geographies of Consumption in the 1990s." *Area* 26 (3): 261–267.
Hagman, O. 2006. "Morning Queues and Parking Problems." *Mobilities* 1: 63–74.
Hamilton-Baillie, B. 2008. "Shared Space: Reconciling People, Places and Traffic." *Built Environment* 34 (2): 161–181.
Hannam, K., M. Sheller, and J. Urry. 2006. "Editorial: Mobilities, Immobilities and Moorings." *Mobilities* 1: 1–22.
Harvey, D. 1989. *The Condition of Postmodernity*. Oxford: Blackwell.
Harvey, P. 2012. "The Topological Quality of Infrastructural Relation: An Ethnographic Approach." *Theory, Culture & Society* 29: 76–92.
Harvey, P., and H. Knox. 2012. "The Enchantments of Infrastructure." *Mobilities* 7: 521–536.
Henderson, J. 2009. "The Spaces of Parking: Mapping the Politics of Mobility in San Francisco." *Antipode* 41: 70–91.
Henley, S. 2007. *The Architecture of Parking*. London: Thames and Hudson.
Home Office 2012. *Protecting Crowded Places*. London: HM Government.
Honoré, C. 2004. *In Praise of Slow*. London: Orion.
Hooker, C. 2007. "Charlie Hooker." In *This Will Not Happen Without You: From the Collective Archive of the Basement Group, Projects UK and Locus+ (1977–2007)*, edited by R. Grayson, 48–49. Sunderland: The University of Sunderland Press.
Jakle, J. A., and K. A. Sculle. 2004. *Lots of Parking*. London: University of Virginia Press.
Jones, R., J. Pykett, and M. Whitehead. 2013. *Changing Behaviours*. Cheltenham: Edward Elgar.
Kodransky, M., and G. Hermann. 2011. *Europe's Parking U-Turn*. New York: Institute for Transportation and Development Policy.
Larsen, J. 2015. "Bicycle Parking and Locking: Ethnography of Designs and Practices." *Mobilities*. doi:10.1080/17450101.2014.993534.
Latour, B. 1993. *We Have Never Been Modern*. Hemel Hempstead: Harvester Wheatsheaf.
Laurier, E., H. Lorimer, B. Brown, O. Jones, O. Juhlin, A. Noble, M. Perry, et al. 2008. "Driving and 'Passengering': Notes on the Ordinary Organization of Car Travel." *Mobilities* 3: 1–23.
Lefebvre, H. 1991. *The Production of Space*. Oxford: Blackwell.
Lipset, D., and R. Handler, eds. 2014. *Vehicles*. Oxford: Berghahn.
Lumsden, K. 2013. *Boy Racer Culture*. London: Routledge.
Manns, J. 2010. "Park-and-Ride: Politics, Policy and Planning." *Town & Country Planning* 79 (3): 144–148.

Marples, E. 1963. "Foreword." In *Parking – The Next Stage*, edited by Ministry of Transport, 4–5. London: HMSO.
Marusek, S. 2012. *Politics of Parking*. Farnham: Ashgate.
Massey, D. 1991. "A Global Sense of Place." *Marxism Today*, June 24–29.
Massey, D. 2000. "Travelling Thoughts." In *Without Guarantees*, edited by P. Gilroy, L. Grossberg, and A. McRobbie, 225–232. London: Verso.
Massey, D. 2005. *For Space*. London: Sage.
McCluskey, J. 1987. *Parking*. London: E & FN Spon.
Merriman, P. 2007. *Driving Spaces*. Oxford: Wiley-Blackwell.
Merriman, P. 2009. "Automobility and the Geographies of the Car." *Geography Compass* 3: 586–599.
Merriman, P. 2011. "Enfolding and Gathering the Landscape." In *Routes, Roads and Landscapes*, edited by M. Hvattum, B. Brenna, B. Elvebakk, and J. Kampevold Larsen, 213–226. Farnham: Ashgate
Merriman, P. 2012. *Mobility, Space and Culture*. London: Routledge.
Merriman, P. 2014. "Rethinking Mobile Methods." *Mobilities* 9 (2): 167–187.
Middleton, M. 1962. "Clutter in the Streets." *Design* 166: 40–49.
Ministry of Transport. 1963. *Parking – The Next Stage*. London: HMSO.
Ministry of Transport and Civil Aviation. 1956. *Parking Survey of Inner London (Interim Report, September 1956)*. London: HMSO.
Ministry of Transport and Civil Aviation. 1958. *Parking Survey of Inner London (Final Report, November 1957)*. London: HMSO.
Morrison, K., and J. Minnis. 2012. *Carscapes*. London: Yale University Press.
Mumford, L. 1964. *The Highway and the City*. London: Secker and Warburg.
Nairn, I. 1955. "Outrage." *The Architectural Review* 117: 363–460.
Newman, O. 1973. *Defensible Space*. London: The Architectural Press.
Nottingham City Council. 2011. *Workplace Parking Levy: Employer Handbook*. Nottingham: Nottingham City Council.
O'Connell, S. 1998. *The Car and British Society*. Manchester, NH: Manchester University Press.
Packer, J. 2008. *Mobility without Mayhem*. London: Duke University Press.
Parkhurst, G. 1995. "Park and Ride: Could it Lead to an Increase in Car Traffic?" *Transport Policy* 2 (1): 15–23.
Parr, M. 2003. Speaking in: *Imagine ... the World According to Martin Parr*, produced and directed by Rebecca Frayn and BBC Television.
Parr, M. 2007. *Parking Spaces*. London: Chris Boot.
Parr, M. 2014. "Parking Spaces Book Description." http://www.martinparr.com/books.
Paterson, M. 2007. *Automobile Politics*. Cambridge: Cambridge University Press.
Pearson, M., and M. Shanks. 2013. "Autosuggestion: A Work of Theatre/Archaeology." Performance Studies International. http://documents.stanford.edu/MichaelShanks/544.
Richards, J. M. 1958. "Man in a Hot Tin Box." *The Architectural Review* 123: 288–301.
Rosa, H., and W. Scheuerman, eds. 2009. *High-Speed Society*. Philadelphia: The Pennsylvania University Press.
Sachs, W. 1992. *For Love of the Automobile: Looking Back into the History of our Desires*. Oxford: University of California Press.
Seiler, C. 2008. *Republic of Drivers*. Chicago, IL: Chicago University Press.
Sheller, M. 2014. "The New Mobilities Paradigm for a Live Sociology." *Current Sociology* 62: 789–811.
Shoup, D. 2006. "Cruising for Parking." *Transport Policy* 13: 479–486.
Shoup, D. 2011. *The High Cost of Parking*. Chicago, IL: American Planning Association.
Smith, C. 2010. *London: Garden City?* London: London Wildlife Trust.
Spinney, J. 2006. "A Place of Sense: A Kinaesthetic Ethnography of Cyclists on Mont Ventoux." *Environment and Planning D: Society and Space* 24: 709–732.
Stewart, K. 2014. "Road Registers." *Cultural Geographies* 21: 549–563.
Thacker, A. 2003. *Moving through Modernity*. Manchester, NH: Manchester University Press.
The Times. 1968a. "Clash Over Multi-story Car Park Near Cathedral." *The Times*, 21 February: 4.
The Times. 1968b. "Car Park Go-ahead for Canterbury." *The Times*, 9 April: 2.
Tomlinson, J. 2007. *The Culture of Speed*. London: Sage.
Tsamboulas, D. A. 2001. "Parking Fare Thresholds: A Policy Tool." *Transport Policy* 8: 115–124.
University of Nottingham. 2013. "University Car Parking Charges Approved Scheme." http://www.nottingham.ac.uk/estates/documents/security/approvedcarparkingschemesummary.pdf.

Urry, J. 2003. *Global Complexity*. Cambridge: Polity.
Urry, J. 2007. *Mobilities*. Cambridge: Polity.
Valentine, G. 1989. "The Geography of Women's Fear." *Area* 21: 385–390.
Wall, G. 2011. "Environmental Parking Charging Policies: A Case Study of Winchester." *Local Economy* 26 (4): 246–259.
Warhurst, J. R., K. E. Parks, L. McCulloch, and M. D. Hudson. 2014. "Front Gardens to Car Parks: Changes in Garden Permeability and Effects on Flood Regulation." *Science of the Total Environment* 485–486: 329–339.
Watts, L. 2008. "The Art and Craft of Train Travel." *Social and Cultural Geography* 9: 711–726.
Whitley, K., and C. Stark. 2014. "Orchestral Manoeuvres in the Car Park." *The Guardian*, online music blog, Accessed June 20. http://www.theguardian.com/music/musicblog/2014/jun/20/peckham-car-park-multi-story-orchestra-sibelius
Wollen, P., and J. Kerr. 2002. *Autopia*. London: Reaktion.

Mobilities and Urban Encounters in Public Places in the Age of Locative Media. Seams, Folds, and Encounters with 'Pseudonymous Strangers'

CHRISTIAN LICOPPE

CNRS, Télécom ParisTech, Université Paris-Saclay, Paris, France

ABSTRACT *This paper provides an analytical framework in which to understand the particulars of digital mobilities, that is the experience and conduct of proximity – or location – aware connected bodies on the move. I argue that the locative media which promotes such forms of awareness make available two versions of the surrounding world, one embodied, one on screen, which articulate with the same 'here and now,' so that a crucial feature of user experience is to recognize and match entities appearing in both versions. The corresponding set of practices may be glossed as a situated 'unfolding.' Finally, I will apply this understanding to discuss future mobilities in urban public places and show how the spread of locative media promotes encounters between 'pseudonymous strangers,' in lieu of the encounters between anonymous strangers which interactionist urban studies deemed crucial to the twentieth-century western metropolitan experience.*

The deafening street around me roared.
Tall, slim, in deep mourning, majestic grief,
A woman passed, lifting and swinging
With a pompous gesture the hem and flounces of her skirt,

Swift and noble, with statuesque limb.
As for me, I drank, twitching like a crazy man,
From her eye, livid sky where the hurricane is born,
The softness that fascinates and the pleasure that kills,

A lightning flash … then night! O fleeting beauty,
By whose glance I was suddenly reborn,
Shall I see you again only in eternity?

Somewhere else, way too far from here! Too late! Perhaps never!
For I do not know where you flee, you don't know where I go,
O you whom I would have loved, O you who knew it!

Charles Baudelaire, 'To a passerby'. Translation by William Aggeler, *The Flowers of Evil* (Fresno, CA: Academy Library Guild, 1954)

1. Introduction

The 'mobility turn' has made us reconsider mobility as more than mere transportation (a means to an end), that is as an activity per se (Sheller and Urry 2006; Urry 2000). In transportation, persons are taken as mobile bodies going from one site to another without being affected by the change in environment, whereas in mobility, persons, understood as uncompleted 'stories-so-far' (Massey 2005), behave as wayfarers, who are constantly transformed by the paths they cut and the ecologies they encounter along the way (Ingold 2000, 2011). However, from its inception, the mobilities turn encompassed two different stances, a conceptual one (reframing mobility as a meaningful activity per se) and a historical one, suggesting that the experience of being on the move was actually transforming along the same lines. I want to focus here on one part of this transformation, which we may describe as 'digital mobilities.' Digital mobilities account for the fact that mobile bodies will increasingly be connected bodies. Mobile digital connectivity allows getting access to any kind of information on the move, but I will focus here on a special case, typical of locative media, that is situations in which mobile bodies are able to connect digitally to surrounding people, places, and things, and in a way which change dynamically in the course of their mobility.

It has been difficult to appreciate the potential contribution of locative media to future digital mobilities for two reasons. First, locative media applications are very diverse today, with the kind of location awareness or proximity awareness they provide varying greatly in its explicitness from one application to another: mobile games, location-sensitive recommendation applications, location-sensitive social network, mobile dating or dynamic car sharing applications. Second, existing locative media have either failed or succeeded mostly in niche markets, which makes difficult to see what might happen should their use become pervasive. Mobile locative media have been making slow and irregular headway in our everyday practices as shown by communication research (De Souza e Silva and Sheller 2014; De Souza e Silva and Sutko 2009; Farman 2013; Hjorth and Richardson 2014; Wilken and Goggin 2012), and Human-Computer-Interaction (HCI) research (Benford et al. 2003; Brown and Laurier 2012).

I will try to overcome these difficulties by drawing on a rich basis of empirical studies to isolate a common feature of locative media, which is to provide a layered experience of mobility, in which the connected mobile individual may experience people, places, and things around her with her own sensorium, but also through the mediation of screen-based connected mobile devices and applications. Seen from a user-centered perspective, mobile locative media can be described as content and applications which are made available to users on mobile terminals, and which provide them with some digital awareness of their surroundings, either through some sorts of real-time digital maps, or through direct notifications that some persons or things are close by. The common and constitutive feature of mobile locative media, irrespective of their technical or functional differences, is therefore that they enable situations in which users have two different kinds of access to what is 'here and there.' Surroundings become simultaneously accessible through the embodied sensorium and through the mobile terminal, with both versions being somehow referred to an ego-centered 'here and now,' so that the issue of correspondence between both experiences of the surrounding world becomes salient. Current approaches of the mobile media experience as 'seamful' (Bell and Dourish 2007; Chalmers and MacColl 2003) and as enacting 'hybrid ecologies' (Crabtree and Rodden 2008) capture rather well the way in which the connected smartphone user is confronted with multiple involvements in a given situation. However, it fails to account for the particulars of the locative media experience in which these involvements rest on different ways of accessing the same 'here and now.' In that respect, the situation of the lost person unfolding his/her paper map to make sense of where he/she is (Brown and Laurier 2005) prefigures the use of mobile locative media, even though there is nothing digital to it. The details of the relevant practices may vary with the details of technologies and settings of use, but I will argue in the first part of the paper that in all cases, they can be glossed as 'unfolding practices': making relevant another version of the surroundings, recognizing entities and matching them across real-life and onscreen experiences, managing mismatches and performing the relevant alignment work.

Locative media have profound consequences with respect to mobility. I will focus here on one aspect, that is the way we meet strangers in urban public places in the course of our mobilities. This is significant, for this a cornerstone of urban sociology, and of the way we understand mobile lives in the modern metropolis. Urban studies have characterized the metropolis as a nexus for large-scale fluxes of strangers in motion. Such a perspective stresses fleeting encounters with passing strangers as characteristic and even constitutive of the twentieth-century metropolitan experience (Goffman 1963; Lofland 1998; Whyte 1980): unlike villages where the default expectation is that we will meet familiar faces as we move along, and in which encounters with strangers are both noticeable and potentially laden with alterity, the default expectation in modern cities is that we will mostly encounter unknown anonymous strangers, mostly similar to us, as we move in public places. This leads to a characteristic form of behavior, adapted to the management of such unfocused encounters with strangers in public, i.e. 'civil inattention' (Goffman 1972), with the gaze as the main resource, and a moral orientation towards preserving the 'negative face' of strangers (Brown and Levinson 1987) and the impersonality of the encounter.

I will discuss how mobile locative media promote a different type of encounter in which users become aware of the proximity or location of other persons nearby, which they mostly don't know and would be unable to recognize by sight. However, these are not complete strangers, as in Goffman's metropolis: mobile applications also often make accessible pictures or profiles of these nearby strangers to which they provide onscreen access, as well as electronic chat resources to possibly interact with them. We will call such encounters, encounters with 'pseudonymous strangers.' Persons nearby are made available under two different guises, as the anonymous stranger we might glance at, and as a digital persona accessible through the proximity-aware locative mobile app, so that real-life identification ('is this person here the one I see on screen, and with whom I may have already chatted?') and matching ('how does he/she correspond to or differ from his/her online profile or from the impression I previously built from our digital exchanges?') become relevant and salient concerns. Because of this duality, and the inherent foldedness it entails, encounters with pseudonymous strangers may lead to original behavior, such as the 'timid encounters' I analyze below. Though encounters with pseudonymous strangers predate the use of mobile locative media – one can think here for instance of people meeting for the first time after seeing a newspaper personal ad (Cocks 2009), prior to the development of mobile locative media, such encounters had to be planned. Location-based mobile applications are extremely powerful resources for multiplying the occasions of randomly encountering pseudonymous strangers. Extrapolating to a future in which most urban denizens would be connected users of mobile locative media, the augmented city becomes the sites of ceaseless encounters with pseudonymous strangers. The default expectation of urban sociality would then become that one would encounter pseudonymous and 'clickable' strangers, not just anonymous strangers as before.

2. Empirical Basis and Methods

To address these questions, I will base the analysis on examples taken from the fast-growing state of the art in this domain, and my own extensive fieldwork, considering in more detail three types of mobile locative media. The first set of applications we will discuss are location-aware games, from early experiments (Benford et al. 2003) to current and emergent large-scale platforms of location-based gaming (Hjorth and Richardson 2014), such as the Google-based Ingress (Morel 2014), with a particular focus on the pioneering Mogi game in Japan (Licoppe and Inada 2006, 2010). I will also consider a proximity-based gameplay in which, unlike in the fully location-aware games which provide dynamic onscreen maps of the current gameplay, players only become aware of the proximity of one another through notifications of their Wi-Fi- or Bluetooth-sensitive terminal, such as the Dragonquest 9 game (Licoppe and Inada 2015). Second, we will take the example of social networking applications, which combine checking in one's location and various form of spatial recommendations, from the insights provided by the seminal Dodgeball (Humphreys 2007, 2010) to current mobile social networking platforms and particularly Foursquare (Frith 2013; Frith, 2014; Licoppe and Legout, 2014). Third and last, we will discuss recent forms of proximity-sensitive dating applications such as Grindr in the gay community (Blackwell, Birnholtz, and Abbott 2014; Brubaker, Anani, and Campbell 2014), and the emergent Tindr and Momo. In the case of Grindr, proximity awareness seems to be used in a way which promotes casual hook-ups (Race 2014), and therefore occasions many sexually oriented encounters with pseudonymous strangers (Licoppe, Rivière, and Morel 2015). Though we will not discuss all experiences of locative media for lack of space, omitting for instance some art-oriented experiments (Southern 2012), this extensive empirical spectrum is enough to provide a broad enough basis on which to draw generalizable conclusions across the field of mobile locative media, which is the purpose of this paper.

This understanding of the salient features of digital mobilities shows that their study unavoidably involves the development of fine-grained research methods, in line with some current concerns regarding mobility research (Buscher, Urry, and Witschger 2010). Since much of the experience of digital mobilities involves coordinating one's on-screen and off-screen changing experiences of one's surroundings, one need to account or record events in both these semiotic fields (see Licoppe and Figeac 2014 for an example). The advent of digital mobilities requires a more pronounced ethnographic turn for mobility research.

3. Using Mobile Devices and Locative Media on the Move: Seams and Folds

The development of smartphones and more generally portable/wearable communication has led to an increasing entanglement of mobilities and communicative practices. An immediate consequence of this, which was noted as soon as mobile phone devices appeared, is that connected mobile users have to articulate their embodied involvement with their urban surroundings with whatever courses of action are active on their mobile terminal and screens. A first conceptualization of these phenomena relied on two notions introduced in the field of HCI: 'seams,' and 'hybrid ecologies.'

'Seams' were initially introduced to revise a fundamental tenet of ubiquitous computing, which had been turned into a design principle, namely that the idea that well-designed technological devices should be made 'invisible,' so that they blend fluidly and smoothly into the different ways users go about their everyday businesses (Weiser 1991). Influential researchers later argued that instead of the smooth and transparent ecology dreamed of by the tenants of ubiquitous computing, users' environments and experiences were unavoidably messy, crisscrossed with varied heterogeneous infrastructures and more or less (mis)matched arrays of resources, and that such messiness had to be addressed head on by design (Bell and Dourish 2007). Seams were thought to occur precisely at such heterogeneous junctures. Seamfulness had to be treated as an integral part of the user's experience, and as a positive resource for users and designers alike (Chalmers and Galani 2004). If we try to describe seams more precisely in pragmatic terms, one could say that in a seamful experience, the local ecology is experienced as offering different opportunities for action relevant to different domains of activity and supported by different locally relevant socio-technical infrastructures. Seams can be contrasted in that respect to boundaries and frames, which are designed and meant to separate 'sharply' categories of persons from domains of conduct. A seamful experience involves the 'simultaneous' relevance of different types of conduct.

Since screens frame and highlight a particular perceptual domain as relevant for users (Introna and Ilharco 2006), inhabiting ecologies where screens and communication technologies in general abound provides for an inherently seamful user experience, that is managing multiple streams of activity both off screen and on screen. While seamfulness describes as well the experience of the connected computer user at home and the smartphone user on the move, connected mobility experiences makes seamfulness particularly salient, for they have to manage simultaneously the people and things they encounter in the course of their mobility (practically, but also morally, for particular normative standards bear on such encounters, as the work of Goffman has extensively shown; Goffman 1963, 1972), and whatever kind of absorptive activity may be relevant at the time. Connected mobile users therefore experience their environment as inherently seamful: the availability of mobile services on portable terminals make simultaneously relevant to them embodied engagements with their surroundings 'in the real life' and engagements with remote persons and events that are made available through the mobile phone. The term 'hybrid ecology' was coined to capture this seamfulness of the connected mobile user's experience. It was initially used in the HCI field to describe the ways in which the players of a precursor and experimental locative-aware game 'exploit hybrid ecological arrangements to achieve cooperative interaction across the physical-digital divide': 'hybrid ecologies are new classes of digital ecology that merge multiple environments, physical and digital together' (Crabtree and Rodden 2008). Seamfulness and hybrid ecologies allow a reconceptualization of both situated user experience and technology design: user experience is characterized by the need to manage smoothly on-screen and off-screen engagements, and design is to enable users to manage smoothly multiple streams of activity in all kind of situations, and particularly mobile ones. These concepts have proven useful in describing the use of mobile media, such as gaming, branching into experiences such as 'ambient play' (Hjorth and Richardson 2014). This kind of situation also shows how seams are woven into situations and not just simply features in the fabric of the material environment (as a seam in a bit of clothing for instance): let the urban pedestrian with phone turn off his/her device and the seamfulness of the situation disappears.

Seamfulness and hybrid ecologies are concepts which are oriented towards the existence of alternative possibilities for action, some of them being mediated by technologies. Seamful situations therefore always involve some degree of equivocality, particularly from the perspective of the other. Is this stranger on the sidewalk mostly engaged in his/her phone conversation, and so to be treated as a 'phoner' or is he/she paying attention to the pedestrian traffic so that he/she may be treated as a competent walker? This also means that users engaged in such seamful experiences must manage that equivocality so as to produce their conduct as intelligible, witnessable and accountable. Seamful situations also require specific skills with which to manage the demands of different forms of engagement. It is not surprising that the use of portable digital devices in many environments has raised many concerns in terms of propriety, particularly with respect to normative expectations regarding proper involvements, but also in terms of risk and safety in urban public settings and traffic.

Seamful situations allow some amount of creativity in user experience, and this is the main reason why they interested designers, attracted as the latter were to what they perceived as the largely untapped innovative potential of seams: users are able to creatively explore the opportunities they offer, while designers are able to create or highlight these possibilities. Seams would therefore become a focus for design (Chalmers and MacColl 2003). Unsurprisingly, the use of smartphones has been regarded as a particularly promising site for seamful design (Barkhuus and Polichar 2010). I want to focus now on a particular subset of connected mobile technologies, in which the service provided by the technology is sensitive in some way to the location of the user and shaped by it, whether it be a GPS-based mapping application, a location-sensitive recommendation service such as Yelp, or a game or social networking application which lets the user see other users in her vicinity. I want to argue here that while still relevant in a broad sense, the concepts of seamfulness and hybrid ecologies do not capture the finer and essential features of the user experience in the case of location-sensitive mobile technologies, and have to be refined.

However, the concepts of seamfulness and hybrid ecology are less relevant to locative media in particular than to mobile media in general. We need to refine these ideas to make sense of the particulars of the mobile locative media experience. With locative media, what goes on on the phone or on screen is reflexively and almost systematically tied to the 'here and now' of the user, as for instance when users' locations are figured on a digital map evolving in real time, and replete with on-screen representations of surrounding persons, things, and places. We therefore propose to speak here of 'reflexive hybrid ecologies' to cover this more specific situation, where the digital experience and the live embodied experience of place are reflexively tied through proximity or location awareness. Situations of use then involve two senses of place, two ways in which to be aware of the location or proximity of other entities: either through one's 'direct' perception through one's usual senses (that is not mediated by any technological prosthesis), or in a way which is mediated either by a digitally induced sensitivity to proximal entities (in the case of proximity-sensitive mobile applications) or by the design of screen-based dynamic representations of the environment (in the case of location-aware applications). These two ways to be sensitive to the spatio-temporal organization of one's environment are deictically articulated: they are both intelligible with respect to the same point of origin, the 'here and now' of the user. This is the constitutive property of locative media in general, framed from a user experience perspective.

This might already occasionally occur as an emergent feature of the use of mobile technologies which do not rely on some form of location awareness in a systematic way, as for instance with an urban commuter using her mobile phone to talk to remote parties while on the move. Though most of the time her conversation will be unrelated to the urban environment in which she moves around, it may happen in special occasions that both get to be related, for instance in situations where the phone conversation is about guiding her to a particular place. Then the talk will involve descriptions of local features, and part of the activity of the person on her mobile phone will be to consider both the environment as she experiences it through her embodied mobility and as it is referred to on the phone, in order to manage some measure of correspondence. Then, in these special circumstances, the mobile phone happens to be used as a mobile locative medium.

Locative media turn what might occur as an occasional feature of mobile phone use into a systematic feature of user experience. Locative media provide some access to surrounding people, places, and information in a way that is systematically and dynamically, though more or less explicitly, shaped by technologically mediated spatial knowledge of the user's location, in an absolute or a relational way (as in the case of proximity awareness). The locative media user experience is therefore inherently seamful (in that respect, it is just one more case of smartphone use on the move), but with a distinctive and fundamental twist. With locative media, two different ways of experiencing place, relevant to the same 'here and now,' are like 'folded' together and unfolded when they become simultaneously relevant. There are two ways for users to access their surroundings: an embodied one, through their unequipped sensorium, and another through the smartphone screen and the digitally mediated version of the world it provides, which is also deictically related to the 'here and now' of the user in some way, whether the relevant service involves making her aware of her location on a map, recommending places nearby, or making her aware of 'interesting' people in her surroundings. Using locative media makes potentially available two deictically articulated versions of the surrounding world at the same time. The experience of mobile locative media as such therefore involves a set of situated practices we could gloss more generally as an 'unfolding' (which also nicely refers back to the older situation of the lost traveler 'unfolding' her paper map to make sense of her whereabouts) in which both ego-centered modes of apprehending one's surrounding are oriented to and made relevant at the same time. Such 'unfoldings' make salient occurrences that are recognizable as potential mismatches: user involvement then usually entails some specific alignment work both to ensure that these two different ways of inhabiting, perceiving, representing, and categorizing their immediate surroundings coincide enough for all practical purposes, and to remedy what might be perceived as consequential discrepancies.

The concept of 'unfolding' allows us to bring together a variety of related situated practices which become apparent in the course of using varied locative media in action. It also highlights a constitutive feature of the locative media user experience, i.e. making recognizable and remarkable any mismatch between the 'here-and-now' as experienced through one's sensorium, and through the mediation of the technology. Attending to such spatio-temporal congruence troubles is a recurrent feature of mobilities assisted by locative media. Once understood as meaningful, such practices can be analyzed empirically in all their meaningful details. Barry Brown and Eric Laurier thus provide a nice and finely argued example of the 'normal, natural troubles of driving with GPS' and of managing such alignment issues (Brown and Laurier 2012). GPS as such will not interest us here, for, though dynamic, it does not (yet) give a sense of the location of other persons or virtual objects. We will rather look here at some locative media which do, and focus on the issue of urban mobilities and more specifically how the pervasive use of locative media might reshape encounters in public places, the study of which has been a cornerstone of urban sociology and urban studies.

I will proceed in three steps. First, I will try to describe the particular forms which such encounters may take when they involve users of locative media and situations which are 'folded,' as described above. Second, we will use this understanding to delineate the normative expectations which underlie the accomplishment of such 'augmented encounters'. Last, we will use this understanding to extrapolate, in the spirit of this special issue, about what future urban mobilities would feel like should the use of locative media become a pervasive feature of urban mobilities in smart cities.

4. Locative Media and the Development of Encounters with 'Pseudonymous' Strangers as a Constitutive Form of Future Urban Sociality

The modern, industrialized metropolis of the twentieth century has been described as a 'city of strangers' (Lofland 1998), in which the paths of large-scale fluxes of mobile bodies unknown to one another mingle in the course of their daily mobilities (Whyte 1980). For twentieth-century sociology, one characteristic and constitutive feature of urban mobilities in metropolitan public places was the expectation that encounters in public places would mostly involve unknown and anonymous strangers (which made unplanned encounters with known passers-by an unusual and noticeable event). 'Stranger' does not involve here the kind of alterity implied, for example, in Simmel's early use of the concept, but highlights the fact of being unacquainted with the persons with whom one crosses paths. This was a transformation from the village or small community perspective, in which the default expectation would be the reverse, that is that one would mostly encounter familiar faces (which would make encounters with outsiders highly noticeable). Goffman has analyzed systematically the organization of focused and unfocused encounters between unknown strangers in public places and highlighted the importance of normative expectations regarding the preservation of face and leaving others mostly alone. In that respect, Goffman's civil inattention appears as a moral expectation and a crucial practical resource in the management of encounters with strangers in urban public places. I will examine here what becomes of such mingling among strangers in public places and the ways in which such minimal proximity-based encounters are managed when the strangers in question – all or at least a significant fraction of them – are mobile locative media users.

4.1. Planned Collocated Encounters with Virtual Acquaintances

To analyze such encounters, it is useful to consider first a very particular kind of encounter between unknown parties, namely collocated encounters in which two persons who have never met before (and are therefore usually unable to identify one another) but who know something about one another actually meet. Such encounters involve others which are both strangers and acquaintances of sorts. He/she is a stranger because we have never met him/her in person and we would not recognize one another by sight. He/she is an acquaintance because there has often been some form of communication between us from a distance, even if we have not met in person. In older times, such remote interactions could have been accomplished through written correspondence, but today it would rely on phone or electronic communication. With respect to the distinct criteria of visual familiarity and acquaintance, such a virtual acquaintance (previously unseen, but heard of or talked to) can be contrasted to the 'familiar stranger' (Milgram 1992). The familiar stranger is that unknown person the face of which may become familiar in the course of repeated mobilities and encounters, but with whom interaction remains non-existent or minimal, precluding the sharing of personal information. The virtual acquaintance is visually unfamiliar, but some personal knowledge about her has been made available, either through prior interaction or by references from other parties. This could be some unknown person whom a well-intentioned friend has recommended you meet for a job or any kind of personal service, someone whose ad you answered to exchange some goods, or a virtual relationship constituted online and with whom you eventually arrange a face-to-face meeting.

The experience of meeting face to face such virtual acquaintances used to be a rather rare occurrence. Such encounters would almost always involve careful planning (and therefore the project to meet): since such protagonists have never met in person before, it would take very special circumstances for them to be able to discover they are collocated by chance, without planning for it. A typical early example would be people meeting through personal ads in newspapers and organizing a first date in some public place. This example highlights a typical and central concern of the parties concerning a first face-to-face encounter between virtual acquaintances: recognizing someone one has never met before but already interacted with, among a crowd of strangers, with an element of possible misdirection at play. In the interview excerpt below, a woman who has been using a web-based dating site describes such an encounter, including how she relied on environmental cues to identify and recognize her unknown partner, and how some mismatch between her expectations and his appearance became a significant issue.

Extract from an interview (A is the interviewee, Q the interviewer).

A: So er, him (she smiles), I met him after work, in the Halles. We were supposed to meet, we had made an appointment … we did not know one another so we counted on the fact that we would be the only two people waiting there and …

Q: Was it outside or inside a café?

A: It was outside at the corner of … it was in the Halles at the corner of Place du Chatelet and … ah I was disappointed because I had fantasized him … I said to myself … well I was fantasizing about his first name, I loved his first name, I loved his voice. He had seemed interesting in the small talk we had made on the phone. And er well he did not match what I expected physically.

The common practice of pre-arranging visual cues such as clothing details is a testimony both to the relevance of identification concerns and to the difficulty of managing such identification practically. With respect to the previous section, such special situations can be said to be 'folded' by design. Because a collocated encounter has been scheduled with an unknown acquaintance, one 'knows' that some unmet party may be around, so that one's surroundings, here and now, can be inspected for her visual discovery, while some representations and expectations of that unmet other are available from a prior history of remote earlier interactions. This duality of access and the joint relevance of both direct perception of the other and of what transpired from previous exchanges at a distance is characteristic of folded situations. A consequence of it is a constant orientation, in the course of identification, to try to match what is perceptible to expectations emerging from a prior story of remote interactions or references *in absentia*. It highlights and makes highly noticeable potential mismatches and misalignments, which may threaten the potential unity of the situation and make relevant some alignment work and requalification of expectations of the persons involved.

4.2. Locative Media: Enabling Chance Encounters between Virtual Acquaintances and 'Pseudonymous Strangers'

Because of the practical problems involved in identifying and recognizing one another, meetings with virtual acquaintances could only exceptionally happen by chance. You had in general to know beforehand that such a virtual acquaintance was around to be able to identify her. But with many kinds of locative media, users are provided with a sense of the proximity and location of other users. If you are a locative media user, a central feature of these devices is to allow you to 'discover' the presence or proximity of other users around you, thus allowing for the possibility of chance encounters with them. These encounters may involve virtual acquaintances, but most of them will concern what we might call 'pseudonymous strangers': these are other users with whom one may never have interacted or talked about before (unlike virtual acquaintances), but who are not complete strangers either, for the locative application usually makes available some info about them, such as an electronic tag name (which may be an actual name or an electronic pseudonym), together with some elements of profile and prior history of use (if privacy settings allows for it). This may happen as well with proximity-sensitive mobile games such as Dragonquest 9 or Ingress, location-sensitive mobile social networking applications such as the general purpose Foursquare, or the dating-oriented Grindr and Tinder.

With such devices, and even more so as they become more pervasive, chance encounters with virtual acquaintances and pseudonymous strangers become not only possible, but also commonplace. With proximity-aware and location-aware locative media, when two users of the same mobile platform get close to one another, their mobile devices may 'recognize' one another, so that the mobile application makes each one aware of the presence of the other under the guise of her digital persona (how they appear within the mobile application), either through notifications or real-time maps. If they choose not to ignore their mutual awareness of the presence of the other and initiate an encounter, they find themselves in the situation of recognizing visually someone they have never met physically, but of whom they have some digital notion (through profiles in the mobile application, and/or prior communication).

Grindr, for instance, is organized around proximity awareness. If one connects to the application, gay contacts and their profiles become available on screen, and they are ranked according to spatial distance, with the closest first. Encounters with potential new partners (or 'casual hook-ups') may now occur by chance, unlike the online dating example above. In the excerpt below, one user describes such an encounter, in which he connects to the application in a gay bar and 'discovers' another Grindr user, previously unknown to him, in the same place:

> Once I happened to see a guy near me in a bar, and between what he emanated and what he had put on his profile, I found that completely paradoxical and I told him so. He looked much better on the profile than he actually looked. But it was not a question of his physical appearance, but more of what he was emanating. It was (on the profile) like 'open guy, kind and positive' and here was a guy who was scowling and did not appear to be easygoing at all. I sent him a message, he looked down and he did not even answer it. I was almost in front of him and he saw very well, and he did not have the balls to reply, even on Grindr. (C. 40 years old)

The main difference between this example and the previous one taken from an online dating interview is that, while in the earlier example, the face-to-face encounter between pseudonymous strangers was the product of design (it was pre-arranged), here it happens by chance. Otherwise, one finds again the characteristic features of encounters between virtual acquaintances and pseudonymous strangers. The 'discovered' potential partner is available and 'known' on screen through his mobile application identifier and profile, so that the match between the digital impression which is made through the mobile application and his physical appearance becomes enough a concern to be mentioned explicitly. The situation involves managing the articulation of the online profile and the co-present person. However, the Grindr user also describes how the situation was managed by using its seamfulness as a resource. Though face-to-face conversation was a possibility (since they had recognized one another), contact was made by text messaging, so that face-to-face interaction appeared to be missing or 'avoided.' Implicit here is the normative orientation that mutual awareness of co-proximity should be acknowledged in interaction through overt recognition and greetings (Licoppe and Inada 2010).

Locative media therefore make possible the 'discovery' of a pseudonymous stranger with whom one has some sort of online connection but has not met in person before, and may not even have interacted online before (as was the case with virtual acquaintances), as an unplanned and serendipitous occurrence. This is a general feature, which is not limited to dating applications. The next excerpt is taken from a field study of Foursquare users in Paris (Licoppe and Legout 2014). In this location-aware mobile social networking application where users may check into particular 'places' (as defined by geolocation and online descriptors), two Foursquare users who have been checking in and competing for the 'mayorship' of the same hotel thus describe their encounter.

> This summer I was having a holiday in La Réunion, and I spent one afternoon with an Air France stewardess, precisely because she was checking in on Foursquare, and she came to look for me at the reception of the hotel, saying 'Are you the Sandrine who just robbed me of my position of mayor for this hotel?' I said yes and we spent the afternoon talking about her job and many other things. And so these are real encounters I would not have made, she would not have talked to me if I had not been on Foursquare. (Sandrine, 45 years old)

This is another instance of a chance encounter between pseudonymous strangers. They had never met, and knew one another only through their having previously checked in at the same Foursquare 'place,' that is an indirect form of interaction. Mutual awareness occasions an encounter. Unfolding the articulation between physical and online co-presence is limited here to recognition and initiation of talk (through direct queries). Matching online profiles and physical appearances is much less of an issue than with dating applications. The way the event above is told, there is also less playful exploitation of the seamfulness of the situation than in the dating examples. It is evoked in the initiation of talk, though, for the opening conversational gambit was supported by a reference to digital awareness (performed through checking in and competing for mayorship).

4.3. Some Characteristic Trajectories in the Unfolding of Encounters with Pseudonymous Strangers: 'Timid Encounters'

Though these chance encounters between pseudonymous strangers may unfold relatively straightforwardly in face-to-face interaction, participants are very much open to the interactional possibilities offered by the seamfulness of encounters unfolding in reflexive hybrid ecologies, as was already the case in the Grindr example above. For instance, the players of the proximity game Dragonquest 9 (a game played on mobile devices in which notifications occur and game actions become possible when players get within a few meters of one another) thus distinguish between 'brave encounters' and 'timid encounters.' In 'brave encounters,' a bit like in the Foursquare example where the air stewardess made herself known to the other Foursquare user competing with her for the title of hotel mayor, players who become aware that they are close to other players they have never met (and only 'know' as characters in the game) make themselves known to them and initiate a face-to-face conversation (Licoppe and Inada 2015). In 'timid encounters,' the same player will stay nearby but will not make himself/herself known in person. He/she will try to act so as to be recognizable as a passer-by while interacting with other players in her vicinity only through the gameplay. 'Timid encounters' exploit the 'seamfulness,' foldedness, and the 'evidential boundaries' (Goffman 1974) of the situation (bodies are visually available to people around but on-screen actions are not). The Grindr encounter in a bar described above was also a timid encounter of sort. Both Grindr users possibly recognized one another as such but elected not to acknowledge it by acting upon it, and kept their interaction within the confines of the mobile screen. In a 'timid encounter,' the timid party deliberately tries to elude visual recognition or avoid acknowledging explicitly while interacting with the other party on screen.

'Timid encounters' provide evidence for the fundamental categorical ambiguity of pseudonymous encounters with strangers in public places that are usually populated by a lot of pedestrians and a few other users of the same location-aware applications (therefore dwelling in the 'same' reflexive hybrid ecology). Two sets of categories are always available to these location-aware, connected urban denizens when their paths cross: that of ordinary mobile strangers, making relevant civil inattention in Goffman's sense, and that of connected users of the same mobile application, which makes them aware of one another's proximity under a digital guise, and which projects the possibility of a more focused encounter and of exchanges through talk or digital resources. 'Timid encounters' play on such an ambiguity, which is also constitutive of other forms of conduct in pseudonymous encounters.

This also opens the way to various types of potentially transgressive behavior. A mild one would be a kind of mobile digital voyeurism, in which one would check the online profiles of other people around without making them aware that she is doing it. A future connected poet might be tempted to check the profile of the beautiful and unknown passer-by who strikes his eye, and would probably not describe such an encounter in the same way as Baudelaire did. For the same reason, encounters with pseudonymous strangers entail specific forms of vulnerability. In the following example, taken from our early study of the location-aware game Mogi, in which players have geo-located maps figuring their location and that of virtual objects to be collected in the city (Licoppe and Inada 2006), one female player, *C*, initiates contact with another player on the basis of their mutual proximity, as perceived on screen (message 1):

1. C : (19:18:20) : 今、近くに居るんだね。

   Now we are close aren't we? ((smiling smiley))

2. D (19:22:36) : 近いねもしかして…!見えたかも?

   yes, very close ((surprised smiley)) I may …have seen you ?

3. C (19:24:54) : 見えてた？

   You saw me?

4. D (19:26:17) : ふふふ

   He He He ((smiling smiley)) ((big heart and small one))

5. C (19:27:54) : 何なに？

   What? what?

In his reply (message 2), the second player, *D*, acknowledges their proximity and hints that he might be seeing the first player. *C*'s next message makes it clear that she has not seen him. This makes salient a visual asymmetry in the situation, the realization and treatment of which rests on its seamfulness (they can exchange text messages independently of face-to-face interaction) and raises a potential issue from her perspective, with respect to the unity and coherence of her situation. She sees *D* on screen and they have acknowledged their mutual proximity, but she cannot reconcile this mediated representation with her co-present experience in which she cannot make him. This is a characteristic issue which may often arise in 'folded' settings, which involve a parallel orientation to two different modes of perceiving the 'here and now' and possible disjunctions between onscreen and co-present experiences. Such concerns are brought to the fore in the following messages, in which *D* first hints at the power such an asymmetry might be giving him (the 'superior' laughter in message 4), thus breaking the normative expectations of equal rights in encounters between strangers in public places, and then evokes the potential disquiet and emotions such a situation might elicit in her.

6. D (19:29:00) : おっ

Oh ((astonished smiley))

7. C (19:32:30) : どうしたの？

What is happening to you?

8. D (19:34:23) : ドキドキしてる……?

Is your heart beating?

9. C (19:34:54) : うん！

Yes!

10. D (19:36:18) : 会う？

Do you want us to meet?

11. C (19:37:46) : ううん。今仕事中だもん！駄目だよ。

No I am at work we cannot ((winking smiley))

His final proposal to meet, eventually declined, can be seen in this sequential context as an offer to resolve the fold-related tension by finally realigning their co-present and online awareness and interactional statuses and making them coincide.

In another telling example from the Mogi field studies, one female player at home expresses her worry to other players with whom she is familiar about a player who often appears on screen near her home and who, although they are spatially close and mutually aware of the proximity, does not try to initiate contact with her

(Licoppe and Inada 2010). Because of such a pregnant 'silence,' he cuts an equivocal figure. Because he keeps silent, he behaves as a stranger would. However, because he is in the game, he is expected to behave as a player (i.e. to acknowledge onscreen proximity by initiating a text message contact), for whom civil inattention is not an option when very close on screen. Anthropology has shown us that things which cross categorical boundaries are often perceived as dangerous (Douglas 1966), and this categorical ambiguity makes this player's conduct equivocal and transgressive in her view, and leads her to perceive him as potentially dangerous, as a possible 'stalker' (a category redefined in this context to describe a player who appears repeatedly close on screen while remaining 'silent'), and to seek the help of her player friends. Playing Goffman's silent stranger within the location-aware game world may occasion disquiet in a way that is understandable by all players. The particular form of seamfulness which characterizes the above situation, and which makes relevant different categorizations and types of related conduct at the same time, shapes the way pseudonymous encounters may become the locus of forms of behavior perceivable as transgressive or threatening.

Chance encounters with pseudonymous strangers are therefore a consequence and a central feature of the use of locative media in urban environments, and part of what makes such an experience distinctive. Their organization, the kind of social conduct that is expected, and the way they might develop in a creative or transgressive way, all build on the seamfulness of the situation (separate mediations for action and communication, 'evidential boundaries,' the relevance of different categorization devices and category-bound conduct, such as passer-by/civil inattention and mobile application user/mobile conversationalist) and on the form of the situation which is made relevant (which makes some configurations recognizable as mismatches with respect to the 'direct' and screen-mediated access to the surrounding world, and the management of such mismatches relevant issues). Some of the behaviors I have described, such as 'timid encounters,' also rest on the fact that there is usually only a minority of locative media users who move along the streets with mobile terminals sensitive to one another's proximity and digital aura. In such urban configurations, chance encounters with pseudonymous strangers are both significant but rarer than the kind of encounters between (unconnected) strangers, the interaction order of which Goffman has described so well. However, if we imagine a future in which the use of locative media becomes so commonplace that all urban denizens are digitally connected and location-aware, then the city becomes less a 'place of strangers' than a place of 'pseudonymous strangers.' As I have tried to show, while Goffman's analytical approach still holds, the kind of interaction order which characterizes them is different from the one he observed in the mid-twentieth-century western metropolis. 'Civil inattention' might still be somewhat relevant, for noticing every onscreen proximity might prove tedious, but it would be a different sort of civil inattention, involving both the minimal gaze management of embodied proximities and a refraining from checking online available information about passer-by and acting upon it. Urban social experiences would necessarily look a bit different if everyone in public spaces carried about them a digital aura which all could perceive through some sort of digital location awareness.

5. Conclusion

We have provided here a framework for making sense of the specificity of the experience of using varied locative media and used it to get a sense of future digital mobilities. While the notion of seams enabled a grasp of situations in which a person might vacillate between various recognizable involvements enacting different on-screen and off-screen mobile activities, I have argued that the use of mobile locative media retained the seamfulness of hybrid ecologies, but with a distinctive twist. Because of screen-mediated proximity – or location – awareness, two different versions of the same 'here-and-now' surroundings are simultaneously available. Using locative media (from the paper map to GPS or proximity-aware mobile applications such as Grindr) involves a set of actions we have glossed as 'unfoldings' in which both deictically articulated ways of considering one's surroundings are brought together in a way which makes potential mismatches or discrepancies salient and therefore makes relevant some sort of alignment work, the exact nature of which will depend on the setting and current activity.

I have used this analytical framework to understand some of the forms of mobility and sociality which could be deemed as characteristic of future digital mobilities, i.e. urban environments characterized by the pervasive use of mobile locative media. I have argued that digital mobile strangers encounter one another in the city as pseudonymous strangers, rather than as anonymous strangers, as was mostly the case in Goffman's metropolis. Pseudonymous strangers are people we have never seen but to the online persona of which we are connected to through our mobile devices, and the proximity of whom mobile locative media make us aware of. Encounters with pseudonymous strangers in urban public places do not just involve a mobile body, quick glances, and civil inattention as in Goffman's description of encounters between anonymous strangers in public places. They make relevant issues of visual recognition and identification, and matching between what is visually available physically and what is available through mobile applications, all sorts of issues which pertain to 'unfolding.' Such person-related forms of unfolding offer specific opportunities for creative behavior (as with the emergence of 'timid encounters' in which one person will elude such visual recognition and matching work while still interacting online) and the recognition of a potential threat, such as the redefinition of a connected other who often lurks nearby, but never proposes any form of interaction as a stalker.

Location- and proximity-sensitive technologies make possible the constant digital discoveries of such 'others' in one's vicinity. In the limit where such technologies become pervasive, all urban denizens are connected to such applications, the default expectations becomes that one meet others as pseudonymous strangers rather than anonymous strangers in urban public spaces. With the development of proximity-aware digital mobilities, the concerns pseudonymous encounters with strangers raise will therefore become central in urban sociality. In a city of pseudonymous strangers, some of the more emblematic figures and characteristic conducts of the nineteenth and twentieth-century western metropolis with respect to encountering strangers become irrelevant. Imagine crossing the path of Baudelaire's *passante* in a fully connected world, where her profile would be available on screen, or trying to recover the transgressive detachment of the *flâneur* while still being connected, and able to click on some profile to view digital info related to pedestrians nearby. The management of engagement and disengagement and even what counts as detachment, and the right to preserve one's tranquility in urban public spaces which were characteristic of the modern metropolis will be deeply affected and reshaped by the development of digital mobilities.

References

Barkhuus, L., and V. Polichar. 2010. "Empowerment through Seamfulness: Smart Phones in Everyday Life." *Personal and Ubiquitous Computing* 15 (6): 629–639.
Bell, G., and P. Dourish. 2007. "Yesterday's Tomorrows: Notes on Ubiquitous Computing's Dominant Vision." *Personal and Ubiquitous Computing* 11 (2): 133–143.
Benford, S., R. Anastasi, M. Flintham, A. Drozd, A. Crabtree, C. Greenhalgh, N. Tandavanitj, M. Adams, and J. Row-Farr. 2003. "Coping with Uncertainty in a Location-based Game." *Pervasive Computing* 3: 34–41.
Blackwell, C., J. Birnholtz, and C. Abbott. 2014. "Seeing and Being Seen: Co-situation and Impression Formation using Grindr, a Location-aware Gay Dating App." *New Media and Society* 17 (7): 1117–1136. doi:10.1177/1461444814521595.
Brown, B., and E. Laurier. 2005. "Maps and Journeys: An Ethno-methodological Investigation." *Cartographica* 40 (3): 17–33.
Brown, B., and E. Laurier. 2012. "The Normal Natural Troubles of Driving with GPS." In *Proceedings of the SIGCHI Conference on Human Factors in Computing Systems*, 1620–1630. New York: ACM Press.
Brown, P., and S. Levinson. 1987. *Politeness: Some Universals in Language Usage*. Cambridge: Cambridge University Press.
Brubaker, J., M. Anani, and K. Campbell. 2014. "Departing Glances: A Sociotechnical Account of 'Leaving' Grindr." *New Media and Society*. doi:10.1177/1461444814542311.
Buscher, M., J. Urry, and K. Witschger. 2010. *Mobile Methods*. London: Routledge.
Chalmers, M., and A. Galani. 2004. "Seamful Interweaving: Heterogeneity in the Theory and Design of Interactive Systems." *Proc. ACM DIS 2004*, 243–252. New York: ACM Press.
Chalmers, M., and I. MacColl. 2003. "Seamful and Seamless Design in Ubiquitous Computing." *Personal and Ubiquitous Computing* 8: 19–30.
Cocks, H. G. 2009. *Classified. The History of the Personal Column*. London: Random House.
Crabtree, A., and T. Rodden. 2008. "Hybrid Ecologies: Understanding Cooperative Interaction in Emerging Physical-digital Environments." *Personal and Ubiquitous Computing* 12 (7): 481–493.
De Souza e Silva, A., and M. Sheller. 2014. *Local and Mobile: Linking Mobilities, Mobile Communication and Locative Media*. London: Routledge.
De Souza e Silva, A., and D. Sutko. 2009. *Digital Cityscapes*. New York: Peter Lang.
Douglas, M. 1966. *Purity and Danger*. London: Routledge.
Farman, J. 2013. *The Mobile Story: Narrative Practices with Locative Technologies*. New York: Routledge.
Frith, J. 2013. "Turning Life into a Game: Foursquare, Gamification, and Personal Mobility." *Mobile Media and Communication* 1 (2): 248–262.
Frith, J. 2014. "Communicating Through Location. The Understood Meaning of the Foursquare Check-in." *Journal of Computer-Mediated Communication* 19 (4): 890–905.
Goffman, E. 1963. *Behavior in Public Places*. New York: The Free Press.
Goffman, E. 1972. *Relations in Public*. New York: Penguin.
Goffman, E. 1974. *Frame Analysis. An Essay on the Organization of Experience*. New York: Harper and Row.
Hjorth, L., and I. Richardson. 2014. *Gaming in Social, Locative & Mobile Media*. Basingstoke: Palgrave Macmillan.
Humphreys, L. 2007. "Mobile Social Networks and Social Practice: A Case Study of Dodgeball." *Journal of Computer-Mediated Communication* 13 (1): 341–360.
Humphreys, L. 2010. "Mobile Social Networks and Urban Public Space." *New Media and Society* 12 (5): 763–778.
Ingold, T. 2000. *The Perception of the Environment. Essays in Livelihood, Dwelling and Skill*. London: Routledge.
Ingold, T. 2011. *Being Alive Essays on Movement, Knowledge and Description*. London: Routledge.
Introna, L., and F. Ilharco. 2006. "On the Meaning of Screens: Towards a Phenomenological Account of Screenness." *Human Studies* 29 (1): 59–76.
Licoppe, C., and Figeac, J. 2014. "Direct Video Observation of the Uses of Smartphone on the Move. Reconceptualizing Mobile Multi-activity". In *Mobility and Locative Media. Mobile Communication in Hybrid Spaces*, edited by A. de Souza e Silva and M. Sheller, 48–64. London: Routledge.

Licoppe, C., and Y. Inada. 2006. "Emergent Uses of a Location Aware Multiplayer Game: The Interactional Consequences of Mediated Encounters." *Mobilities* 1 (1): 39–61.

Licoppe, C., and Y. Inada. 2010. "Locative Media and Cultures of Mediated Proximity: The Case of the Mogi Game Location-aware Community." *Environment and Planning D: Society and Space* 28 (4): 691–709.

Licoppe, C., and Y. Inada. 2015. "Mobility and Sociality in Proximity-sensitive Digital Urban Ecologies: 'Timid Encounters' and 'Seam-sensitive Walks'." *Mobilities*. doi:10.1080/17450101.2014.988530.

Licoppe, C., and M.-C. Legout. 2014. "Living Inside Mobile Social Information: The Pragmatics of Foursquare Notifications." In *Living Inside Mobile Information*, edited by J. Katz, 109–130. Dayton, OH: Greyden Press.

Licoppe, C., C.-A. Rivière, and J. Morel. 2015. "Grindr Casual Hook-ups as Interactional Achievements." *New Media and Society*. doi: 10.1177/1461444815589702.

Lofland, L. 1998. *The Public Realm Exploring the City's Quintessential Social Territory*. New York: Aldine de Gruyter.

Massey, D. 2005. *For Space*. London: Sage.

Milgram, S. 1992. "The Familiar Stranger: An Aspect of Urban Anonymity." In *The Individual in a Social World: Essays and Experiments*, edited by John Sabini and Maury Silver, 51–53. New York: McGraw-Hill.

Morel, J. 2014. "Ingress. Mobilités et sociabilités dans un jeu de réalité augmentée [Mobilities and Socialities in an Augmented Reality Game]." *Interfaces numériques* 3 (3): 447–472.

Race, K. 2014. "Speculative Pragmatism and Intimate Arrangements: Online Hook-up Devices in Gay Life." *Culture, Health & Sexuality: An International Journal for Research, Intervention and Care* 17 (4): 496–511. doi:10.1080/13691058.2014.930181.

Sheller, M., and J. Urry. 2006. "The New Mobilities Paradigm." *Environment and Planning A* 38 (2): 207–226.

Southern, J. 2012. "Co-mobility: How Proximity and Distance Travel Together in Locative Media." *Canadian Journal of Communication* 37 (1): 75–91.

Urry, J. 2000. *Sociology Beyond Societies: Mobilities for the Twenty-first Century*. London: Routledge.

Weiser, M. 1991. "The Computer for the 21st Century." *Scientific American* 265 (3): 94–104.

Whyte, W. H. 1980. *The Social Life of Small Urban Spaces*. Ann Arbor: Edwards Brothers Publishers.

Wilken, R., and G. Goggin. 2012. *Mobile Technology and Place*. New York: Routledge.

Mediated Pedestrian Mobility: Walking and the Map App

ERIC LAURIER*, BARRY BROWN** & MOIRA MCGREGOR**

**School of GeoSciences, Institute of Geography, University of Edinburgh, Edinburgh, UK*
***Mobile Life Centre, Kista, Sweden*

ABSTRACT *While walking has always been mediated, the arrival of smartphones with multiple apps has changed how we walk and how we use apps. In this study, we investigate the relationships of pedestrian-in-the-street and app-user-on-screen actions. We display and describe a series of intersubjective practices constituted by, and with, walking while using a mobile device. The video data used are from a larger study of pedestrians using smartphones in urban settings, with our analysis here turning on how a smartphone is used and interacted around to accomplish walking together. Our approach draws upon ethnomethodological conversation analytic studies of the sequential and category-based organisation of mobile and on-screen actions. The analysis shows how walking actions (such as unilateral-stopping, turning, restarting) are connected to map actions such as displaying the map, manipulating the scale and monitoring the movement of the you-are-here dot. We conclude with remarks on the collaborative inter-subjective nature of walking with apps.*

Introduction

Pedestrians do not just walk. They rush, they dawdle, they stroll, they amble, they circle, they pause, they stop, they edge past, they saunter, they plod, they advance, they retreat, they backtrack, they lead, they follow. Walking happens as a host of more prepositional, intentional and consequential actions: they walk towards, they walk away, they walk off, they walk into. Nor do pedestrians make naked contact with the places where they are walking: hiking shoes protect their feet while hill-walking, price tags and labels shape their trajectory while shopping, podcasts envelope them in comic dialogues while they walk to work.

In this article, we pursue and extend, then, the idea of that there is no such thing as walking-in-itself. Our goal is not to produce 'the differential configurations of self and landscape' (Wylie 2005, 336) but rather to move towards the intersubjective production of mediated mobile practices. Our practices of walking in the city are not all the same, nor are they the same as walking barefoot in the park. They are mediated by the ancient technologies of shoes, pavements and pedestrian crossings. We walk around the supermarket with a shopping list (Cochoy 2008), stand at a bus stop reading the timetable (Watson 2009), move around art galleries in relation to an exhibit (vom Lehn 2013) and stand for a while in the street pausing to listen to a busker (Smith, forthcoming). How we move through places on foot has been further transformed by the arrival of various new technological systems and their related media. The smartphone looms large as a device that we walk with, one that introduces multiple kinds of media into pedestrian practices. One of the significant 'new-yet-old' media that it has brought to walkers is the map. The map app departs from pedestrians' pre-existing paper maps in several ways: map apps have a 'you-are-here' dot and provide suggested routes between points, map apps can be rescaled and rotated to align with the compass orientation of the device and they integrate with online search engines. They allow their users to drop pins and other markers into the map to show them where things are in relation to their current location.

Walking is seldom somnambulistic: to be moving your feet at all is also to be taken to have a reason for doing so. When the customer walks up to the counter, they are taken to be about to pay for their goods. Walking, then, not only glosses over the myriad forms of pedestrian movement, it glosses over the intended actions that constitute pedestrian practices. Austin's (1962) 'How to do things with words' was a landmark lecture series in how it explored language not as a referential system for making true/false statements, but as something we *do* things with. Research on mobility has built upon Austin and others' work in considering the performative role of mobility yet has missed Austin's careful differentiation of what specific ordinary words do. For instance, in his discussions of truth statements, he examines the varying effects if three words that are commonly used in those statements (e.g. 'entails', 'implies' and 'presupposes'). In this article, we deepen Mobility Studies' engagement with Austin's work by considering how pedestrians *do things with walking* while never letting go of how *walking is done with things*.

In what follows, we will examine how the practice of walking is central to reading the map app and how the reading of the map app is reflexively, if asymmetrically, tied to the walking. Through examining pedestrians walking together, we will explore how walking plays its part in making recognisable to one another what is happening on a moment-by-moment basis, whether each and other knows where they are going and where each and other will go next. Specific features of walking actions such as pace, trajectory, halts, turns and so on provide resources for making sense of the map together, and showing the sense we are making of the map since pedestrian movement is itself inscribed into the map. GPS technology places a 'you-are-here' dot on the map, it scrolls the map and it is drawn upon to establish the scale. As Frith (2012) puts it, the map apps allow an inversion of the classic state cartography. Rather than the states predetermining how the world will appear on the map, the users of map apps, in interactions with their algorithms and interfaces, establish the centre points, scales and layers of their own maps, and they do this on a moment-by-moment basis.

Doing Things with Walking

While Ryave and Schenkein (1974) opened their classic study of 'doing walking' with a brief mention of 'crawling, hopping, running, cartwheeling, jumping, skipping' (1974, 265), they settled on navigation and the reflexive production and recognition of pedestrians walking together as their main focus (Ryave and Schenkein 1974). We can, however, garner hints of the actions produced by walking in their remarks on how single persons carefully manage their movement on the pavement to avoid being seen as doing activities like 'following' or 'joining'. Their work was picked up almost two decades later by Lee and Watson's (1993) studies of pedestrians in street markets, boarding trains and navigating crowded stairways. Lee and Watson described how walkers, by their general positioning in relation to one another and relevant features of the environment, produce 'flow files' and undertake pathfinding. Moreover, they demonstrated how walking in public produces a moral order manifest in the accountability of pace and preferences for minimising disruption to movement (Lee and Watson 1993).

It is only recently that these early studies have been returned to, in part as a response to the rise of mobility studies and how mobility then features in embodied practices (see the collections (Haddington, Mondada, and Nevile 2013; McIlvenny, Broth, and Haddington 2014). Growing out of ethnomethodology and conversations analysis (EMCA), the contemporary studies have closely examined how talk, gestures, objects and environments are interwoven with courses of action. This turn towards mobility has been through undertaking studies of the vehicular, or, to put it another way, how people do things with, and through, moving as a vehicle. The vehicular units in question might be pedestrians entering a building (Weilenmann, Normark, and Laurier 2014), a couple travelling by car (Mondada 2012b), a small aircraft landing (Nevile 2004) or a couple doing Lindy Hop (Broth and Keevallik 2014). Across these varied settings, the studies have been concerned with, firstly, how people jointly coordinate and accomplish their mobility on a moment-by-moment basis and, secondly, how they do this *as* particular categories of social unit.

In terms of the moment-by-moment organisation of action, the classic ethnomethodological questions of 'why that now?' and 'what next?' have, in a number of these studies of movement, been supplemented with 'why that *here*?' and '*where* next'? For the vehicular unit in motion, by their very movement through the environment, they have a constantly changing 'here' which then also provides changing access to events and to perspectives. Motion also shifts the proximity and distance of the features of the environment, for driving: side streets, exits, entrances (Haddington and Keisanen 2009) or, for supermarket shopping: aisles, freezer units, bakery counters and bargain baskets. Concomitantly by their ongoing movement, each unit has changes in their sequential geography of where they have just been and could possibly go next.

Sensitive to the situated nature of walking as action, EMCA studies have examined how the category of the unit itself produces distinctive ways of walking while also being reflexively reconfigured by those mobilities. In considering the social categories of the mobile unit we are returned to the premise of theories of walking and just as there is nothing that is walking-in-itself, walkers are never just walkers (Broth and Mondada 2013; Mondada 2012a; de Stefani and Mondada 2014). It is not just that walkers are themselves different categories of group (e.g. traffic wardens, office workers, museum visitors) – they undertake varied practices that then shift their categorisation, so that in the case of museum visitors, they may shift from touring the exhibits to browsing in the museum shop and then on to standing together in the museum restaurant queue. By doing so, they shift from producing, being recognised and responded to as visitors, to shoppers and then to diners.

The recent EMCA studies have shifted the emphasis from the early studies' interest in the ongoing maintenance of walking together to the various reconfigurations that happen when groups do things like 'moving off' or 'moving on' (Broth and Lundström 2013; vom Lehn 2013; Mondada 2012a). The ongoing configuration of the relative positions of group members and their talk mutually elaborate one another so that, for example, in de Stefani's (2013) study of supermarket shopping, one member of a joint unit can announce that he is going to get the bread by himself, begin to depart and thereby temporarily break the pair into two solo shoppers. What de Stefani highlights is that this is *joint* mobile action and that rather than the one who has departed being able to thereby do what he intended or not, next actions are oriented towards and being responded to by the other party in parallel or subsequently. In de Stefani's case, when one shopper departs, the other shopper does not stand still, but instead 'walks beside' without a word uttered. An incipient departure for choosing bread alone is cancelled by a walking action that means the bread will be chosen together. Through studies of couples and friends shopping together, it is not that walking actions need to be infinitely varied, de Stefani shows how such simple mobile actions as 'stop' and 'go' are central to the mutual intelligibility of shopping as a pedestrian vehicle. In this article, we draw upon and extend these EMCA studies to consider how walking actions are used by wayfinding tourists in relation to mobile media.

Mediated Walking

There is a long history of the mediation between the walker and the ground, most obviously through boots and shoes but also through tracks, roads and pavements (Ingold 2004; Michael 2000). For the walker in the city, not only is the very ground smoothed and shaped to allow them to look up from their footing (Ogborn 1998), but also they are surrounded by a multitude of urban signage and advertising, as sketched out by Latour (2003) in his study of Paris. In it, he documents a city bristling with street signs, optical devices, receipt slips, guidebooks and more; Latour's meditation on the mediation of Paris preceded the widespread adoption of the smartphone by urban pedestrians. The phone has brought to our lives in public spaces, new forms of encounter and new forms of co-presence (Licoppe 2013). It has facilitated game-playing on the move and thus brought a changed way of playing together (Richardson 2013), some games making 'remote' others co-present with the pedestrian (Southern 2012). Map apps on smartphones arrive on the shoulders of a longer pedestrianisation and transformation of media, introduced through the use of new mobile devices. The transfer of telephony to the walker through the mobile phone brought its own changes to the assumed location of both the caller and the called (Ling 2004; Weilenmann 2003). Mobile phones also brought SMS text messaging and consequently new practices of writing and reading as forms of being together with remote others (Harper et al. 2005).

Walking through crowded urban spaces (or equally through empty rural spaces) has become hybridised with a series of practices previously found in other spaces: listening to music, gossiping with remote friends or ordering tonight's dinner (Frith

2012). As the smartphone has enfolded further digital media (e.g. Google search, Wikipedia, YouTube, yelp), it has brought ever more practices to pavements: searching, checking recommendations, playing games and, as we focus on in this article, following maps (Brown, McGregor, and McMillan 2015; de Souza e Silva and Sutko 2011). The very walking of the vehicular unit is registered in the media by the system, a delegation of the work that a diligent map reader might have undertaken with pencils, compass, street signs and landmarks. Meantime, the mobile device is not only tracked by GPS and made sense of by the system; the walkers take it out of their pocket, raise it close to their eyes, show it to or hide it from other walkers. As we noted above, the device itself has a mobility in relation to the talk, gestures and courses of action of the walkers. Our interest is in map practices now remediated by the map app, the movements of smartphone itself and how map apps are then occasioned in particular ways within the pedestrian practices of wayfinding.

Wayfinding Accomplished through Mediated Walking Practices

As a walking unit, tourists move through the city in distinctive ways while also drawing upon their individual and relative mobility as a resource. For example, while one member of the group stops to consult the guidebook, other members can walk around inspecting the surrounding environment for street names, information signs, landmarks, etc. (Laurier and Brown 2008). By the member's pedestrian practices of 'walking away' from the immobile guidebook-reader, the other members of the group categorise themselves as 'doing reconnaissance'. Having walked away to examine the surrounding environment, there are then expectations that they will return with information relevant to their wayfinding task.

It may not be initially clear the difference that walking as a particular mode and material arrangement of mobility makes to wayfinding with map apps. To turn briefly to GPS navigation systems in cars, while this is obviously a vehicular social unit moving together, the car passengers have fixed relative positions that are not easily interchangeable. The GPS units are usually also fixed to the dashboard as part of the ecology of the car interior. Once car travellers are settled into their seats, they stay in them. From their relative positions, drivers and passengers can gesture with eyes and hands, turn their heads and lean their torsos, but they are otherwise restricted in their mobility in relation to one another (Brown and Laurier 2012; Haddington and Keisanen 2009). Pedestrians using a map app on a smartphone are a different entity because they have distinctive embodied ways of walking together. They can individually 'stop' and 'go', change their relative pace thereby 'falling behind' or 'rushing ahead'; they can adjust the gap between their bodies, they can break apart their vehicle to let other pedestrians through or shift from side by side to single file to pass through narrow gaps or doors (Conein, Félix, and Relieu 2013). The device itself can be moved into and out of pockets and bags, closer and further away from eyes, tilted towards or away from the other walker and so on. With the new resources that map apps provide, they require different sorts of intersubjective mobile inquiry and are still settling into how we wayfind and wayfare. We hope then in what follows there will be some novelty in examining how a particular map app is drawn upon and itself redrawn during practices of walking as tourists in the city.

Methodology

The recent generations of lightweight video cameras developed for extreme sports have leant themselves to changed recording techniques for moving cameras with mobile subjects (see Brown, Dilley, and Marshall 2008; Spinney 2011). Following multiple camera techniques developed by McIlvenny (2015), for recording cyclists, we asked our project participants to each wear cameras that sat on their chests. While for McIlvenny's cycling studies, bicycle frames and riders' helmets provide several points for attaching cameras to then capture forwards, rearwards and sideways angles, we did not want to overburden our participants with cameras. Consequently, each project participant had only one camera which then captured a chest-height view ahead but also allowed us to work out the orientation of their torso, their direction and speed of walking. Depending on their orientation towards one another, it produced recordings of each other's body orientations and gestures from time to time. Other researchers (e.g. Licoppe and Figeac 2013) have used camera glasses that better capture head movements, other gestures and the inspection of smartphones. However, our preference was for the better quality images produced by sports cameras that then also allowed us to feed a lapel mic into them for better sound quality. The major drawback of our set-up is that we seldom see facial expressions and miss a number of gestures, objects and elements of the environment that will have been in the moving visual field of the participants. Not all of these absences are solved by camera glasses, however, because these cameras move with the head rather than the eyes of participants. During our data gathering, we also installed screen capture apps on our participant's own phone or provided them with a project phone with the software already installed.

The recordings from the cameras and screen capture were transcribed as graphic transcripts (Laurier 2014). The format merges elements of the Jeffersonian conventions for transcribing conversations with those of graphic novels. Where speech bubbles overlap, participants' speech is overlapping. Equally where speech bubbles are placed far apart, this indicates gaps between speaking. The temporal precision of textual transcripts is sacrificed in favour of, both the ease of reading of graphic transcripts and, more importantly, the visual details that they allow readers to register without lengthy textual descriptions.

We recruited 15 participants in the cities of Stockholm and London; five of these were individuals and five were couples or friends. Our set-up for gathering data was a hybrid of classic task-based wayfinding experiments and ethnographic studies of map use. The study participants were asked to go on a 'typical day out' which was convergent with what they were planning to do when they were recruited. We collected 24 h of video data that was then edited into smaller fragments around wayfinding events and the use of map apps. The fragment we will draw upon in this article is a specimen, which is rich in material for analysing the interconnection of walking actions and the map app. It involves two American tourists, ('Carol' and 'Dani') realising that they have made a mistake while walking towards a destination for lunch, and then repairing their mistake by consulting the map app.

(1) Walking as Merely Walking and Checking the Route

As we have argued above, when people walk alone or together it involves much more than simply walking together side by side. Questions of where walkers are going next, who knows where the group is going next and whether they are more or less lost, are posed and answered through the combination of talk and walking actions. The production and recognition of one another's walking actions involves adjustments of pace, changes of direction, starting and stopping. In this first section, we will examine the walking actions of the tourists where their pace and direction simply continues the journey itself.

The map app on Carol's smartphone has been used earlier to plan their route, identifying a short list of landmarks and street names. The underground stations along the main street that they are walking past were picked out as landmarks. In passing (or not passing) the landmarks that were earlier identified, the map is then made sense of in a reflexive relationship with the journey that they are making (Liberman 2013). Consulting the map is shaped by the very features of the urban environment that were picked out on each earlier consultation of the map app.

By virtue of having her smartphone out and checking the map app at the outset of their journey, Carol has also become accountable for this planned route. Wayfinding in groups, be they formal like a ship's crew or informal like friends on a day out, involves distributing labour and accountability. Carol has acquired the responsibilities of being the map reader and being accountable for that as part of their journey in search of lunch. However, while these responsibilities remain potentially relevant during the journey, whether the walkers shift in and out of their map reading incumbency depends upon the other things they are doing amidst the walking activity. Walking in the city has a permeable quality because many other activities to seep into it.

In Transcript 1, Carol and Dani are approaching the entrance to an underground railway station. The camera angle forwards from Carol's midriff. The station entrance is on the street corner just beyond the red carpet entrance. Dani is walking on Carol's right-hand side, out of shot.

Closing in on the station entrance coincides with a lull in their conversation. Carol, by shifting the position of her smartphone, initiates a new course of action around the smartphone, which brings back her incumbency as map reader. However,

Transcript 1. Walking together, checking and showing smartphone.

for Dani (on the right), what is being initiated is not immediately apparent, given that there is an array of actions that may be initiated by raising a smartphone – such as answering an incoming call, checking a text message, taking a photo, checking email and, of course, checking the map. In continuing her gesture by then angling the smartphone towards Dani, Carol offers what is on-screen for Dani. She then goes on to say, 'oh we're at this one', the oh-preface marking an epistemic change (e.g. it is a discovery that they are at this underground station). The referent of 'this one' being established by the timing in relation to: what they are currently walking past and her pointing over the screen at a station on the map (see Transcript 1, last two panels). Dani has only minimal involvement, which is unsurprising considering her distance from the screen. Carol had to move the smartphone close to her eyes to inspect the map app closely enough to read road names, etc. By contrast, Dani merely glanced at the smartphone screen when it was tilted towards her in Transcript 1, the distance combined with her glancing indicative of just such minimal involvement.

Throughout this checking of the smartphone, there is no change in the pace or direction of their walking and, in that sense, the journey continues onward without alteration or correction. What Carol's comment, 'so we're close', accomplishes is preparing the pair of them for the more detailed checking and approach-work that will be required on reaching their destination. The walking has shifted from walking past one landmark after another, to walking towards their journey's final destination. In sum, we have seen how the media is drawn upon smoothly to maintain continuous walking while wayfinding work is undertaken. Things, however, do not continue to be so smooth.

(2) Unilateral Stopping

For pedestrians walking together in the city, coming to a halt jointly without a verbal request is part and parcel of walking as a group. Pedestrians stop at the edges of roads when they cannot cross, they stop to take photographs, to let people pass or make their way through entrances and exits (Weilenmann, Normark, and Laurier 2014). Closely coordinated stopping is possible because of the walkers' shared access to the many features of, and events in, the environment that call upon them to pause their journeys or stop travelling entirely. In fact, between Transcriptions 1 and 2a, Carol and Dani have just passed through one of the most common urban features which make stopping relevant: the edge of the pavement where it meets the road.

Carol and Dani are advancing together continuously until Carol unilaterally stops (see Transcript 2a). Stopping walking can be done more or less abruptly and we have examples in our corpus of walking coming to a gradual stop. Here, when Carol stops, she stops sharply. Had she produced her stop more gradually, then Dani might still have stopped in tandem with her. Moreover, she is not at a point where stopping was already spatially relevant (e.g. the earlier edge of a pavement).

However, Carol's stopping is neither quite as abrupt nor unannounced as it might seem from Dani having missed it and walked on for several steps. Having checked her smartphone, Carol did not put it away and was still inspecting it as they crossed the road, having had to restart the maps app (we will look at that in more detail below). Moreover, the stop is pre-figured by an audible sigh. From the sigh, Carol then further prefigures difficulties by an 'uh' which Dani also produces in overlap (panels 2 and 3 of Transcript 2a). It may be that Dani's 'uh' precedes a question in

Transcript 2a. Unilateral and sharp stopping by Carol.

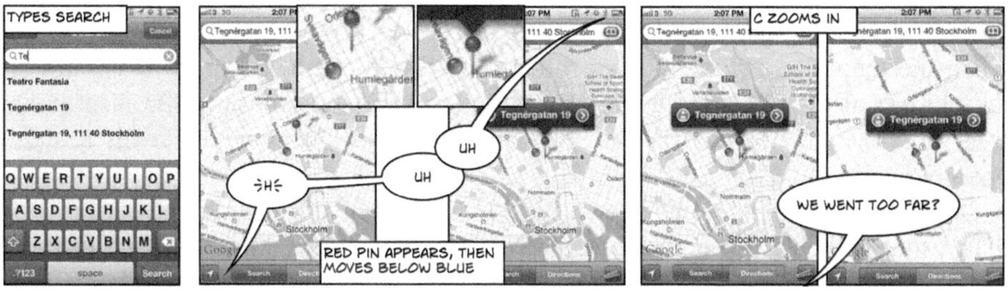

Transcript 2b. The you-are-here and destination dots.

response to the sigh over what Carol's difficulties are. In addition, Carol had mentioned that they were 'close' to their destination and it may be that Dani then anticipates problems around locating their destination given they have never been to it before so do not know what it looks like nor where it is.

Carol begins to account for her unilateral stop by saying 'we went too far?' (panels 3 and 4). The question is an account, which fits nicely with an immediate stop, which could itself precede turning back on their path as a likely next walking action. Carol's querulous tone may mark her own scepticism over what she has seen on the map app given that she had not detected any problems on passing the train station only three minutes earlier. Dani on hearing the question turns tightly around to reorient herself both to a backward direction from their previous ongoing trajectory and to create a face-to-face spatial formation with Carol (see Mondada 2012a on changing formations). While unilateral stopping can accomplish various things, the responses to it are limited. The other walker(s) can also then stop or can continue walking. In de Stefani's (2013) work on supermarket shopping, when one shopper stops to inspect goods, the other can continue ahead or also stop and join them in inspecting their goods. For the wayfinders, when the map reader stops, the other members of the party best not continue given that it is quite whether they should be walking ahead at all, that is in question.

(3) Walking as a You-are-here Dot

Map applications on smartphones ('map apps' such as Google Maps) have brought the you-are-here blue dot into common use. In the process, the previously fixed 'you-are-here' dot now moves across the map as its users move through the environment. Map apps depart from older paper maps in several other ways: they provide maps that can be re-scaled and rotated to align with the compass orientation of the device. We might expect that this combination of the locational resources of the you-are-here dot and the auto-rotation of the map would allow us to easily locate where we are while our journeys progress, given that they do the work of alignment of the map with the environment for us. Yet these very manipulations of media require further sense-making (Laurier and Brown 2008) and the things that we do with maps when we are on the move are much more than locating where we are (Brown and Laurier 2005).

Part of the reason that a unilateral stop occurs is that Dani does not have shared visual access to the map app. Before the unilateral stop, if we look at what was happening on Carol's screen, the map app had restarted, and Carol had retyped the search term for the destination they had agreed upon earlier (panel 1 of Transcript 2b). The map reappears on her map app when they are mounting the pavement. Carol was sighing, perhaps due to having to type their destination for a second time, when the map app restarted. The problem pre-figuring 'uh' occurs when Carol sees the red pin of their destination shift on the map app (see Transcript 2b inset boxes). It is this change of state on the map app that Carol responds to with a sudden unilateral stop.

On each check of the map, Carol checked that their dot has moved further along their route, which is up the yellow street, prominent in the middle of the map. The trouble that Carol detects is that the red destination pin has shifted from being above the you-are-here blue dot, to below it. In response to the trouble, Carol both lifts the device closer and pinches to zoom into the map for more detail. However, as we noted earlier in her comment, 'We went too far?', her tone remains sceptical or in some way uncertain over whether there really is a problem given her process of checking landmarks as they went along. The questioning tone is over the claim made by the map app even while, by being said aloud, it provides access to the mistake that Carol and Dani might have made in their journey.

While the long-standing convention for walking with paper maps has been that we work out landmarks and check them off as we pass them, the map app shows our location with a dot or similar icon. In the map app, the dot itself moves along the route that we have planned, or indeed off the route, depending on whether we deviate from the suggested route. Through its movement, the you-are-here dot is aligned with the direction of our walking. A common trick we found pedestrians using was walking forward slightly to see where the blue dot moved on the map app, as a way of checking on their direction. Or, as is the case here, checking the movement of the dot as part of checking on the physical landmarks that they were passing. There is in the map app then a new set of resources for establishing where we are at any point in our journey.

(4) Reconnaissance and Re-stopping

To summarise our findings so far, we have looked at the smooth progression of walking with a map, the use of a unilateral stop by the map reader that was, in part, generated by having asymmetric access to the smartphone and, further, we considered how the map app is used in making sense of where walkers are. We will now examine the distribution and articulation of wayfinding work in walking actions as each walker moves in relation to another. Remembering that where they are and where they should go next is in question, we see how each party tries to help establish answers to these questions.

Transcript 3a. Dani's reconnaissance and Carol's confirmation of their mistake.

Transcript 3b. Reading the street names and adjusting the map.

If we consider the earlier distribution of the wayfinding work, Dani had been journeying with Carol but only peripherally involved in the wayfinding (e.g. through being shown the device as they passed the train station entrance). After Carol's unilateral stop and account, Dani's involvement increases through an immediate mobile reconfiguration. In the first and second panels of Transcript 3a, Dani pivots rapidly around, continues walking while then also pointing towards the other side of the road. As part of this mobile reconfiguring, she jointly asks a wayfinding-relevant question of Carol as the map reader. The question shows an understanding of Carol's immobility as being momentarily 'stuck'. Dani's pointing gesture is prolonged as she takes a few steps in the new direction that would allow them to backtrack and begin a search for the destination. In short, she is quickly attuned to the problem and seeking to establish the general direction of their destination.

We see the complex projectability of gesture and action that walkers can achieve in Dani's bringing together of two orientations: her direction of walk is in one direction while her pointing is in another. Her walking is towards some other walking-relevant target (e.g. the pedestrian crossing visible in the final frame of Transcript 3a) while her gesture is dealing with where she should look for their destination. What we can detect from Carol's camera movement (see the changing perspective in each panel in Transcript 3a) is that Carol rotates to converge with the direction of Dani's onward movement and by panel 4 appears to have faced towards the pedestrian crossing as well thereby bringing their projected next actions into alignment.

Transcript 4a. Carol restops and continues her map reading.

Transcript 4b. Reading the street names and adjusting the map.

Carol's response is more complex than moving together with Dani to the pedestrian crossing: even though she uses her walking actions to follow those of Dani, she ignores Dani's question 'Is it over on this other side?', instead answering her own earlier self-directed question, 'We went too far?', by providing a self-confirmation that, 'Yeah we went too far'. However, it is not so much that she is ignoring Dani, because whether they have actually overshot is not something that Carol has established. On epistemologically thin ice, Carol is not moving too far toward the street until she has clarified whether there is a problem at all with their wayfinding, or whether the problem is with the map app, or with her current reading of the map app.

This parallel course of involvement continues when, after Carol begins to move she then restops. This second stop, after restarting in the new direction, is accounted for with 'I think' as then being generated by her continuing uncertainty over whether they have overshot (which can be seen on the screen capture in Transcript 3b) and their movement. By zooming in on the map app, Carol has brought up street names and a more detailed view of the city blocks. Dragging the map up slightly (Transcript 4b) seems to help her become certain that they have gone too far rather than that there is some other problem.

What the restop accomplishes, then, is putting their back-tracking on hold while Carol then examines the map app for a second time. Her restop is visible to Dani who then makes a few small steps to become proximate to Carol, showing by her walking an awareness of the need for further scrutiny of the map app. In this close formation Dani, does not simply wait on an answer from Carol reading the map app but makes a different request for a name of a street for Dani herself to look for. In her previous request, she was asking for the area to look towards, now with Carol and Dani having an aligned orientation towards a tableau of shops, cafes and restaurants Dani asks for a name (see similar orientational work in the city street in Mondada 2009). With a name, she could then scan the names of premises in the tableau. Carol holds off Dani's request with a brief 'Wait' (panel 2, Transcript 4a). Carol then raises her own difficulties around their movement (e.g. 'which way') before finishing by becoming more certain again. She does this through repeating the phrase 'we went too far' for a third time so that it shifts from being a sceptical question to an affirmative conclusion.

Aside from the restop putting their progress on hold, it also rearranges the pair's orientation towards one another and to the street features ahead. From the 'hold' position, it is then Carol who restarts their walking and thereby, Carol also shifts from becoming the uncertain map reader to showing she knows where they are and where they are going. In terms of the relationship between walking and who knows what (Mondada 2013):,this brief pause in Carol's walking is used to mark her uncertainty about continuing with the adjustment to their route. Having made sense of the map app in relation to their journey, Carol then makes visible that certainty by 'setting off' with a trajectory that reinitiates the move towards the crossing point.

What this complicated fragment of the interaction shows is how the reading of the media is reshaping their walking and being reshaped by their walking actions, not only moment-by-moment but in parallel courses of action, one of which orients towards inspecting the urban environment and the other from inspecting the media (see also Haddington et al. 2014). Moreover, it underlines how a walking unit can distribute the wayfinding tasks amongst its members, even if, in this case, the party seeking to do some reconnaissance work is not given the information she would need to do so.

(5) Leading and Following

In the previous section, we detailed how checking the map app and the city environment were accomplished through the sequences of walking stop/starts and questions: Dani, becoming the scout, turning around while pointing, Carol, becoming the map reader stopping, turning, then stopping again. We left Carol and Dani at a point where the map reader had become certain from her careful map reading that they have gone 'one street too far'. In this section, we will examine how Carol makes it recognisable that she is now confident about where they are going. The walking action that Carol produces is to shift ahead of Dani's relative position. Rather than create a mobile configuration that is side by side, or single file, she produces a step-ahead (see Transcript 5). The walking unit then becomes organised into one person leading and the other following a step behind. Carol's shift to leading the group is based upon her consultation of the map that she continues to check during her walk towards the pedestrian crossing, (see panel 2 of Transcript 5).

It is hard not to feel that Dani has to some extent been sidelined in the wayfinding task given that she was not supplied with a name and thus was not able to search the neighbouring environment for it (an example of this can be found in Laurier and Brown 2008). However, Dani does not give up and although Carol is pressing ahead Dani then speeds up to close in on her again (see Transcript 6). While attempting to close the gap, Dani pursues the knowledge that will increase her parity in the wayfinding work by shifting from a request for a name to just an orientational question about their destination point on this street. Not only does Dani ask a further

Transcript 5. (From Dani's perspective): Walking ahead of the other.

Transcript 6. Accelerating to recreate a side-by-side.

question in relation to finding their destination, she recreates a side-by-side arrangement of the walking pair so that as a walking unit, the leader-follower arrangement is being ended. By her walking actions, in a reflexive relationship with her requests for information, she is managing her epistemic status in relation to their wayfinding as tourists. It is at this stage that she finally receives an answer to her 'side' question from Carol.

Dani has an oh-prefaced realisation (see panel 3 of Transcript 6), before expressing her misapprehension that their destination was in the immediate area (e.g. 'not *on* this street'). Despite her desire to do so, Dani could not have helped solve their wayfinding problem by looking around for their destination because it was still beyond their visual access on the far side of a city block. What leads Dani to this realisation is hard to determine, but we can speculate that it is in the walking action of Carol. Carol is walking ahead but she is not looking around at names of the places on the other side of the street and she does not point at anything. She only walks towards the other side of the street. In other words, by the absence of the embodied qualities of inspecting from Carol's walking that would show that the destination is close enough to be looked for, Dani is able to establish that their destination is 'not *on* this street'.

In this final section, we examined firstly how the leading and following configuration is assembled by Carol thereby re-establishing her as the map reader certain of where their destination lies. Secondly, we showed how Dani caught her up and re-established a side-by-side walking configuration, which is indeed the format for walking together as acquaintances or friends. Were Dani to have dropped behind Carol it would have been towards showing reluctance, dismay, tiredness or some other problem with their current togetherness as friends exploring a foreign city.

Thirdly, we also touched upon how Carol's manner of walking at this point is made sense of in relation to their search for their destination. To be walking without scrutinising the immediate environment was then analysed by Dani, also in relation to the absence of replies to her requests, as meaning that their destination was further than she had thought. In other words, the journey itself is reflexively tied to expectations around walking actions. When we near a new and unknown destination, we adjust our walking to allow us to find that destination.

Conclusion

If the first studies of the mediation and materiality of walking began, appropriately, where the foot hits the ground, in this article we have ascended to the mid-riff, shoulders and almost to eye level. Ingold's careful consideration of walking showed us the 'bias of the head over heels' (Ingold 2004, 331; see also Michael 2000). In the rise of Modernity, feet were treated as farther and farther from thinking, even as they were augmented with shoes and disciplined through marching and other body techniques. For Michael (2000), the boots are never simple intermediaries between the walker and nature, they also interfere, intervene and influence through pain, style, standardisation and erosion. It would be no surprise to say that the smartphone also interferes, intervenes and influences the walker. In the crowded streets of the city, the eyes that were freed from the ground by shoes and pavements to become Goffman's urban pedestrians that avoid bumping into others by 'scanning for other people' (Ingold 2004, 238) – these eyes are now dragged downward again by the varied activities undertaken on smartphones. The smartphone-pedestrian does not move in the ways of the iPod-orchestrated-walker that Bull (2001, 2005) described so well. By following two tourists for a few minutes while they are wayfinding in the city, what we have described is this evolving mediation of the smartphone and walking that picks up where Ingold and Michaels stop, at the feet of the pedestrian. We have brought thumbs onto map apps, walkers approaching pedestrian crossings, and the mutual and responsive orientation and trajectories of walkers.

Our article brings to their work on the mediation of walking not only this rapidly evolving technology that is the smartphone, it directs our attention to doing mobility together, and to responsive yet also divergent and convergent mobile actions. Where much of the work on walking has examined the link between the perceiving self and perceptible culture and environment (Wylie 2005), we have begun instead with how two persons are inter-subjectively producing their journey in the city. A central problem for pedestrians is around how walking is involved in producing the ongoing sense of: where we have been, where we are now and where we are going next. On collective pedestrian journeys, there are jobs to do that are distributed amongst members of vehicular units, and these then also distribute locally determined accountabilities for onward (and backward) travel and the employment of media.

Through the analysis of video and screen recordings, we have delineated how walking is one part of the gestalt of action which sustains the intelligibility (and unintelligibility) of what the tourists are doing in ongoing real time. We identified a number of walking actions: 'merely walking', unilateral stopping, turning while pointing (which is different from turning to point), restopping, restarting, walking ahead, catching up, walking beside, changing direction. For each action, we described how walking actions were used within the emerging response to, and analysis of, the tourists' problem of having overshot their destination, and in relation to one tourist's analysis and ongoing configuration of the map app. The reflexive relationship between reading the map app and the journey, threaded through courses of action familiar from earlier studies of maps and wayfinding, but the map app brings with it new mediated practices, not least walking-the-dot. The latter may be analysed, as it was here, simply through accumulation of the previous checks of the you-are-here dot on the map or, as it was with other participants in our study, through actually walking forwards while checking to see what direction the blue dot then moved on the map.

In examining the complexities of how smartphones are implicated in pedestrian practices, what we hope is apparent, is that we have only just scratched the surface of the touchscreen's reconfiguration of, and by, mobility. If we consider for a moment, the ways in which smartphones are involved in friends being together and apart, catching buses, game-playing, looking up information, not to forget listening to music and a myriad other practices, there is much for mobility studies to understand about the mediation of not just walking actions but, of course, driving, cycling, travelling by bus, train and plane. We have attempted to move from the classical lone walker, to study a contemporary mode of walking: together with touchscreens.

Acknowledgements

A special debt is owed to Allison and James for their assistance in the revising and redrafting of this paper. To our study participants we also owe a huge thank you.

References

Austin, J. L. 1962. *How to Do Things with Words: W. James Lectures Delivered at Harvard, 1955.* London: Oxford University Press.

Broth, M., and L. Keevallik. 2014. "Getting Ready to Move as a Couple: Accomplishing Mobile Formations in a Dance Class." *Space and Culture* 17 (2): 107–121. doi:10.1177/1206331213508483.

Broth, M., and F. Lundström. 2013. "A Walk on the Pier: Establishing Relevant Places in Mobile Instruction." In *Interaction and Mobility*, edited by P. Haddington, L. Mondada, and M. Nevile, 91–121. Berlin: Walter de Gruyter.

Broth, M., and L. Mondada. 2013. "Walking Away: The Embodied Achievement of Activity Closings in Mobile Interaction." *Journal of Pragmatics* 47 (1): 41–58. doi:10.1016/j.pragma.2012.11.016.

Brown, B., and E. Laurier. 2005. "Maps and Journeys: An Ethno-methodological Investigation." *Cartographica: The International Journal for Geographic Information and Geovisualization* 40 (3): 17–33, UT Press.

Brown, K. M., R. Dilley, and K. Marshall. 2008. "Using a Head-mounted Video Camera to Understand Social Worlds and Experiences." *Sociological Research Online* 13 (6): doi:10.5153/sro.1818.

Brown, B., and E. Laurier. 2012. "The Normal Natural Troubles of Driving with GPS." In *Proceedings of the SIGCHI Conference on Human Factors in Computing Systems,* 1621–1630. New York: ACM.

Brown, B., M. McGregor, and D. McMillan. 2015. "Searchable Objects: Search in Everyday Conversation." In *Proceedings of the 18th ACM Conference on Computer Supported Cooperative Work & Social Computing*, 508–517. New York: ACM. doi:10.1145/2675133.2675206.

Bull, M. 2001. "The World According to Sound Investigating the World of Walkman Users." *New Media & Society* 3 (2): 179–197. doi:10.1177/14614440122226047.

Bull, M. 2005. "No Dead Air! The iPod and the Culture of Mobile Listening." *Leisure Studies* 24 (4): 343–355. doi:10.1080/0261436052000330447.

Cochoy, F. 2008. "Calculation, Qualculation, Calqulation: Shopping Cart Arithmetic, Equipped Cognition and the Clustered Consumer." *Marketing Theory* 8 (1): 15–44.
Conein, B. C. Félix, and M. Relieu. 2013. *The Visual Sense of Togetherness. Passing through Doors.* Presented at MOBSIN 4, Nice: University of Nice, Sophia Antopolis, Copies available from the authors.
Frith, J. 2012. "Splintered Space: Hybrid Spaces and Differential Mobility." *Mobilities* 7 (1): 131–149. doi:10.1080/17450101.2012.631815.
Haddington, P., and T. Keisanen. 2009. "Location, Mobility and the Body as Resources in Selecting a Route." *Journal of Pragmatics* 41 (10): 1938–1961. doi:10.1016/j.pragma.2008.09.018.
Haddington, P., T. Keisanen, L. Mondada, and M. Nevile. 2014. "Towards Multiactivity as a Social and Interactional Phenomenon." In *Multiactivity in Social Interaction: Beyond Multitasking*, edited by P. Haddington, T. Keisanen, L. Mondada, and M. Nevile, 1–30. Amsterdam: John Benjamins.
Haddington, P., L. Mondada, and M. Nevile. 2013. "Being Mobile: Interaction on the Move." In *Interaction and Mobility*, edited by P. Haddington, L. Mondada, and M. Nevile, 3–61. Berlin: Walter de Gruyter.
Harper, R., L. A. Palen, and A. Taylor. 2005. *The Inside Text.* Dordrecht: Springer.
Ingold, T. 2004. "Culture on the Ground the World Perceived Through the Feet." *Journal of Material Culture* 9 (3): 314–340.
Latour, B. 2003. *Paris: Invisible City.* Paris: bruno-latour.fr. Accessed June 12, 2012. http://www.bruno-latour.fr/virtual/EN/index.html
Laurier, E. 2014. "The Graphic Transcript: Poaching Comic Book Grammar for Inscribing the Visual, Spatial and Temporal Aspects of Action." *Geography Compass* 8 (4): 235–248. http://doi.org/10.1111/gec3.12123.
Laurier, E., and B. Brown. 2008. "Rotating Maps and Readers: Praxiological Aspects of Alignment and Orientation." *Transactions of the Institute of British Geographers* 33 (2): 201–216.
Lee, J. R. E., and R. Watson. 1993. *Interaction in Public Space: Final Report to the Plan Urbain.* Paris: Plan Urbain.
vom Lehn, D.. 2013. "Withdrawing from Exhibits: The Interactional Organisation of Museum Visits." In *Interaction and Mobility*, edited by P. Haddington, L. Mondada, and M. Nevile, 65–90. Berlin: Walter de Gruyter.
Liberman, K. 2013. *More Studies in Ethnomethodology.* New York: State University of New York Press.
Licoppe, C. 2013. "Merging Mobile Communication Studies and Urban Research: Mobile Locative Media, 'Onscreen Encounters' and the Reshaping of the Interaction Order in Public Places." *Mobile Media & Communication* 1 (1): 122–128. doi:10.1177/2050157912464488.
Licoppe, C., and J. Figeac. 2013. *Patterns of Gaze Switching in the 'Naturally-Occurring' Uses of Smartphones in Urban Mobile Settings.* Paris: TELECOM ParisTech.
Ling, R. S. 2004. *The Mobile Connection.* San Francisco, CA: Morgan Kaufmann.
McIlvenny, P. 2015. "The Joy of Biking Together: Sharing Everyday Experiences of Vélomobility." *Mobilities* 10 (1): 55–82. doi:10.1080/17450101.2013.844950.
McIlvenny, P., M. Broth, and P. Haddington. 2014. "Moving Together: Mobile Formations in Interaction." *Space and Culture* 17 (2): 104–106. doi:10.1177/1206331213508679.
Michael, M. 2000. "These Boots are Made for Walking: Mundane Technology, the Body and Human-Environment Relations." *Body and Society* 6 (3–4): 107–126. doi:10.1177/1357034X00006003006.
Mondada, L. 2009. "Emergent Focused Interactions in Public Places: A Systematic Analysis of the Multimodal Achievement of a Common Interactional Space." *Journal of Pragmatics* 41 (10): 1977–1997. doi:10.1016/j.pragma.2008.09.019.
Mondada, L. 2012a. "Garden Lessons: Embodied Action and Joint Attention in Extended Sequences." In *Interaction and Everyday Life: Phenomenological and Ethnomethodological Essays in Honor of George Psathas*, edited by H. Nasu and F. C. Waksler, 279–296. Plymouth: Lexington Books.
Mondada, L. 2012b. "Talking and Driving: Multiactivity in the Car." *Semiotica* 191 (1/4): 223–256. doi:10.1515/sem-2012-0062.
Mondada, L. 2013. "Displaying, Contesting and Negotiating Epistemic Authority in Social Interaction: Descriptions and Questions in Guided Visits." *Discourse Studies* 15 (5): 597–626. doi:10.1177/1461445613501577.
Nevile, M. 2004. *Beyond the Black Box.* Aldershot: Ashgate.
Ogborn, M. 1998. *Spaces of Modernity.* New York: Guilford Press.
Richardson, I. 2013. "Touching the Screen: A Phenomenology of Mobile Gaming and the iPhone." In *Media*, edited by S. Mobile, 133–151. London: Routledge.

Ryave, A. L., and J. N. Schenkein. 1974. "Notes on the Art of Walking." In *Ethnomethodology; Selected Readings*, edited by R. Turner, 265–274. Harmondsworth: Penguin.

Smith, T. Forthcoming. "Street Performing Spectacles and Ordinary Life." PhD diss., University of Edinburgh, Edinburgh.

Southern, J. 2012. "Comobility: How Proximity and Distance Travel Together in Locative Media." *Canadian Journal of Communication* 37 (1): 75–91.

de Souza e Silva, A., and D. M. Sutko. 2011. "Theorizing Locative Technologies through Philosophies of the Virtual." *Communication Theory* 21: 23–42.

de Stefani, E. 2013. "The Collaborative Organisation of Next Actions in a Semiotically Rich Environment: Shopping as a Couple." In *Interaction and Mobility*, edited by P. Haddington, L. Mondada, and M. Nevile, 123–151. Berlin: Walter de Gruyter.

de Stefani, E., and L. Mondada. 2014. "Reorganizing Mobile Formations: When 'Guided' Participants Initiate Reorientations in Guided Tours." *Space and Culture* 17 (2): 157–175. doi:10.1177/1206331213508504.

Spinney, J. 2011. "A Chance to Catch a Breath: Using Mobile Video Ethnography in Cycling Research." *Mobilities* 6 (2): 161–182. doi:10.1080/17450101.2011.552771.

Watson, R. 2009. *Analysing Practical and Professional Texts*. Farnham: Ashgate.

Weilenmann, A. 2003. "I Can't Talk Now, I'm in a Fitting Room: Formulating Availability and Location in Mobile-phone Conversations." *Environment and Planning A* 35: 1589–1605.

Weilenmann, A., D. Normark, and E. Laurier. 2014. "Managing Walking Together: The Challenge of Revolving Doors." *Space and Culture* 17 (2): 122–136. doi:10.1177/1206331213508674.

Wylie, J. 2005. "A Single Day's Walking: Narrating Self and Landscape on the South West Coast Path." *Transactions of the Institute of British Geographers* 30 (2): 234–247.

The Anticipated Futures of Space Tourism

MARK R. JOHNSON* & DARYL MARTIN**

**Science & Technology Studies Unit, Department of Sociology, University of York, York, UK*
***Department of Sociology, Wentworth College, University of York, York, UK*

ABSTRACT *This paper examines the development of the 'space tourism' industry, a concept which primarily denotes the development of space technology for recreational or leisure purposes. It will first theoretically locate space tourism with relevant streams of the mobilities literature, primarily research on aeromobilities and tourism mobilities. It will then summarise existing literature on space tourism, focusing especially on the different models of space tourism that have been proposed and the subtle but important differences between them. The analysis then explores the perceptions of feasibility of space tourism from those within the space sector, the anticipated changes to living these forms of space tourism would bring with them, and the comparisons with existing forms of mobility that are drawn by many in the space sector when attempting to 'sell' the value and potential of space tourism. In the final part of the paper, we explore many of the implicit assumptions held by this nascent industry, the uncertain position of the passenger within these conceptions of future space tourism, and identify avenues for future research into this emerging form of mobility.*

Introduction

The space industry is currently developing a number of models for a potential future 'space tourism' business, which primarily denotes the development of space technology for recreational or leisure purposes. Space tourism is but one part of the current privatisation of much of the global space industry. Andre Lebeau (former Director of the European Space Agency) has argued that Cold War competition and bloc prestige has now vanished (Salomon 1996) and space agencies cannot extract the same 'unswerving support' from states they could in the past. He claims that because of this, the balance of power has shifted towards industry, both reducing the role of the state and 'revealing the maturity of space technology' when deployed straight into the market, much as a 1994 *Eurospace* (1994) report argued that the space industry is now fundamentally driven by customer demand. The rhetorical construction by industry figures of space tourists of the future will be a focus of this paper, alongside a reflection on their implied mobilities. Space tourism, we argue, sits within a sector whose representatives seem more comfortable evoking mobility systems and infrastructures of the past when imagining emergent practices, perhaps in part because of the sociopolitical, economic and technological challenges and contexts of the present.

Given the difficulties in reaching orbit for any given space mission, the potential feasibility and realism of space tourism has been widely questioned both within the space sector and in public discourse. These issues will be examined through both prior scholarly works dealing with space tourism and relevant bodies of mobilities research, and via data acquired during the first author's research into the UK space industry. This research involved 26 in-depth semi-structured interviews with individuals employed across a wide variety of space sector roles, many of whom were in very senior positions within public or private bodies. Interviewees came from a range of origins – scientific, commercial, governmental and technological – and both public bodies and private companies. Approximately one-third of the respondents were interviewed at the UK Space Agency, the UK's governmental body for space activities oversight, management and funding, and the remaining two thirds were contacted via the use of snowball sampling across the wider field of the UK space industry. A total of eight interviewees are quoted in this paper, one-third of the total research sample, and all have been assigned anonymous placeholder names. Space tourism was only one topic of several explored in the original interview research. Themes in the interviews were defined by the interviewees, not the researcher, in keeping with a commitment to inductive research in the overall research design. Given the many topics and the use of semi-structured interviews, only in one-third of the interviews did the topic of space tourism arise in sufficient analytic depth to be reflected in this paper.

This paper will first chart recent developments in cognate strands of the mobilities literature (primarily in the area of aeromobilities and tourism mobilities), before offering a summary of the literature on space tourism to date. Here we will draw out the different models of space tourism that have been proposed and the subtle but important differences between them. The analysis will then explore three main areas based on the first author's research: firstly, the perceptions of feasibility of space tourism from those within the space sector; secondly, the anticipated changes to living these forms of space tourism would bring with them; and thirdly, the comparisons with existing forms of mobility that are drawn by many in the space sector when attempting to 'sell' the value and potential of space tourism. It thus analyses the range of perspectives on space tourism put forward by those working within the space sector and critically examines the range of discursive and rhetorical tools used to put forward these predictions of space tourism futures. Space tourism seeks efficient, reliable, cheap and routine launches (Sadeh 2005) and safe and flexible access to space (Launius 2000), both a long way from the predominantly risky, costly and long-term space industry that currently exists (Elhefnawy 2004). This raises a number of questions – What will a space tourism industry actually look like, and how will it feel to the tourist? Who will these tourists be? Will the industry be run by states, a partnership dependent upon or funded by states, or subsist on its own? Most crucially for this paper, what future forms of mobilities do those promoting space tourism aim to create, and how do these actors use existing forms of mobility and transport to discursively position space tourism as a realistic and viable concept?

Cosmopolitan Visions: Aeromobilities, Tourism Mobilities and Space Travel

The last decade has seen the emergence and establishment of a rich research agenda on the mobilities shaping and structuring contemporary societies, enacted through the everyday movement of bodies, objects, images, ideas and capital (Hannam, Sheller, and Urry 2006; Merriman 2015). Topics have ranged in scale from the movement of individuals to the systems in which they move, as well as a burgeoning literature on the use of mobile methods to explore these (Büscher, Urry, and Witchger 2010). In the first issue of *Mobilities*, Adey (2006a, 83) argued for an understanding of mobility as 'profoundly relational and experiential'; put simply, there are differences in how people experience movement, and there are political implications of these differences. In his analysis of contemporary mobilities, Urry (2000, 167) offers an analysis of the significance of the 'growth of post-national citizenship', in which he draws out questions of political belonging and citizenship in societies where individual experiences are increasingly bound up in social processes and interactions that transcend national boundaries. This reconfiguration of the nation-state entails a new way of understanding social dynamics and requires a new analytical imagination when researching contemporary societies, beyond what Beck (2002, 18) termed the 'methodological nationalism' and 'monologic imagination' of traditional social sciences. In their place, Beck argues for a cosmopolitan vision and re-calibration of sociological and political analysis of contemporary social worlds and the routine cultural processes they facilitate; it is worth here briefly reflecting on how scholars have developed ideas of aeromobilities and tourism mobilities, two sets of mundane practices that can illuminate the aspirations and expectations industry figures articulate for future space travel.

The troubling of traditional ideas of the social that are anchored within national boundaries is, in some respects, symptomatic of the way in which technologies have shaped the cultural and political imaginaries associated with mobile practices, not least those of aeromobility. Skrbis and Woodward (2013, 61) locate an 'aesthetic of cosmopolitanism' in the development of global air travel networks, and Urry (2007, 136) argues that 'without the complex systems of air travel, what is now termed "globalization" would be utterly different', highlighting the significance of not only the aeroplanes that move bodies and objects across the globe, but also the institutional and expert systems that synchronise and regulate these movements. For Adey (2010, 21), 'the aeroplane's mobilities threaten the vital integrity of "territory", "border" and "citizen"' in the present day, but also when thinking about its histories of use throughout the twentieth century. The aeroplane's capacity for troubling such boundaries is, in part, associated with the perspective offered through aerial height, a perspective that can be seen as gendered and socially shaped (though see DeLyser [2011] for a discussion of early female pioneers of aviation). The aeroplane has been a key cultural artefact in the construction of narratives of citizenship; Adey (2007, 2008) has persuasively charted the development of the aviation industry and the significance of its (now) everyday spaces, such as the airport and their viewing balconies, in the building of 'air-mindedness' amongst the general population. Air-mindedness, Adey (2006b, 346) suggests, 'can be conceived as a moral geography that worked to define particular ideas, beliefs and behaviours acceptable for one to attain citizenship of the nation, locality or city'. It has evolved throughout the twentieth century, via representations of aviation in education and popular culture, as part of wider belief systems in the capacities of different nation-states to live up technologically to their rhetorical aspirations (whether ideals of liberty in the US, or the state-planned utopias of the USSR).

> Air-mindedness was effectively the organizing complex or ideal that orchestrated the enthusiasm and development for aviation. It distributed politics, technologies, people, education, policies, places and things for the development of flight and the betterment of the nation. (Adey 2006b, 346)

This distribution tied together layered forms of 'visioning' (Adey 2010, 21), creating mechanisms for the governing of populations out of the romance of the aerial perspective, where visions of national identity become intertwined with visions of social control, and achieved through technological innovation. These processes of visioning within aeromobilities, and questions of how the discourses and practices bound up in them play into wider social narratives, are germane when approaching the social and political implications of space travel, and the discursive registers in which it is developed.

Also salient when situating the emergent discourses of space travel is an understanding of contemporary tourism mobilities; tracing how scholars have characterised the movement of tourists across the globe can be instructive in thinking through the ways in which we might come to think about travel outside the earth's orbit, and the social and political implications nested within such practices. Metaphors of vision have been central to the conceptualisation of tourism, from Urry's (1990, 1995) articulation of the 'tourist gaze', and his later, related, observation of the 'aesthetic cosmopolitanism' that connotes contemporary touristic practice, to Germann Molz's (2010, 345) recent characterisation of round-the-world travellers as carrying with them 'a mobile panoptic gaze that takes the world as its object'. Germann Molz's interpretation of contemporary tourism mobilities is based on the observation that many Western travellers are increasingly at home whilst on the move; in this, she builds on Clifford's (1997) metaphor of 'dwelling-in-traveling', troubling conventional distinctions between the affective qualities of home and away, and she shares affinities with Edensor's (2007) argument that tourism is an increasingly mundane practice, troubling traditional understandings of travel as escape from the everyday. Her work is empirically supported by the importance of new mobile communication devices and infrastructures in allowing forms of co-presence with distant others whilst touring the world, to the extent that we should think of tourist practice as 'a complex "assemblage" of bodies, mobilities, portable technologies, concrete infrastructures, networked spaces, and virtual places in which the social and the technological are mutually determined' (Germann Molz and Paris 2015, 176). This interdependence of mobility and technology is a wider theme within the field of tourist mobilities, as identified in Hannam, Butler, and Paris's (2014) recent review, which also places tourism as deeply implicated in formulations of cultural citizenship. Again, the link to cosmopolitan visions is central, with Germann Molz (2008) drawing on Bruce Robbins's definition of 'actually existing cosmopolitanism' within the context of tourism as related to not placelessness, but rather a situated sense of place, which is tied into 'a reality of (re)attachment, multiple attachments, or attachments at a distance' (Robbins 1998, 3). The translation of this sociopolitical dynamic into tourist mobilities, argues Germann Molz (2008, 338), aligns with a deep-seated cultural tendency in touristic practice: the reproduction of 'a privileged cosmopolitan appropriation of the world' through the sense of entitlement, comfort and confidence that certain social groups carry with them in their journeys, wherever these take them. Asking ourselves how the development of space tourism is presented (that is, to whom and by whom) will be important in bringing to the fore questions of the relational aspects of this future mobility system.

Although there has been a relatively slow take-up amongst social scientists on the cultural significance of technologies in the space sector (Parker and Bell 2009), there have been analyses which have placed the space industry within wider narratives of capitalist development (Dickens 2009; Parker 2009) and debates around the changing configurations of the nation state (Collis 2009). It is instructive to think through the emergence of novel forms of mobilities with reference to what we know historically of the development of earlier forms, in order that we diagnose cultural continuities in the adoption of new technologies (Cresswell 2011). To this end, Collis (2009, 62) draws historical parallels in contemporary debates about the legal geographies and ownership claims of the geostationary orbit with territorial disputes during the era of colonial expansion. It is also instructive to draw out the affective strands of how mobilities are experienced and articulated to help compose an understanding that is multiple and politically attuned (Cresswell 2010), even when considering a technology in its nascent state. So it is interesting to reflect that space travel has been characterised in terms of its 'hopeful cultural resonance', as a 'text that blends technology and fiction to produce an image of the future that may well be better than the present' (Parker and Bell 2009, 3), not least because that implies a 'sort of orientation to the future that is hard to sustain, as the ice caps melt and the credit crunch bites' (Parker and Bell 2009, 4).

As Urry (2011, 139–141) has argued, there are different techniques embedded within narrative accounts of socio-technical life in the future, with predictions based on normative visions, prospective models that extrapolate from present trends and the practice of scenario building which distinguishes likely and desirable outcomes. In academic and industry discussions of space travel, these techniques interweave. Credibility is an essential aspect of successful space programmes and presenting a convincing vision and planned process for a space programme is key from the very beginning of a space programme, for futures only exist so long as they are considered realistic and viable (Brown, Rappert, and Webster 2000). A mission that is too far from the established space industry norm at the point of its suggestion may not be seen as a credible prospect. The 'science fiction wonderment' of early spaceflight (Huntley, Bock, and Weingartner 2010) faded to be replaced by much tougher and tighter assessments of risk and reward, and all contemporary space programmes operate within these new restrictions, especially relevant since the task of launching a human 'tourist' – the central goal of private spaceflight (von der Dunk 2011) – into space is generally agreed to be far more difficult than launching a satellite. Moving space technology into the private sector makes it unclear how risk and managerial priorities will be managed in these companies. This leads to present uncertainty about the future prospects of the technology, its economic application and its social implications; and so, it is to the future scenarios of space tourism that the article now turns, before moving on to the reporting of empirical findings.

Anticipated Forms of Space Tourism

Anticipation and prediction have become a crucial aspect of many high-technology industries, not least in light of their enmeshment in processes of environmental change (Dennis and Urry 2009; Urry 2011). The claim that future impacts of a technology can be anticipated reinforces the linear deterministic notion of technological outcomes (Williams 2006) that STS has worked so hard to challenge. Space tourism has seen a number of anticipated forms, which each point towards a different conception of what space tourism may be a decade or more in the future. Not just do these forms propose different outcomes for the space tourism industry, but these different conceptions very clearly imagine different mobility practices, institutional forms and social dynamics that may arise around such an industry. Three different terms have been used in the existing literature (both scholarly and publicity oriented) to describe space tourism, each of which points towards a different future that, although all three are similar, are distinct in terms of the markets they seek to attract and the formulations of space travel they imagine. As well as simply 'space tourism' – though as we will see this sometimes also takes on a specific nuanced meaning, as well as a more generic all-inclusive one – the terms 'citizen space exploration' and 'personal spaceflight' have also been used in the context of explaining the goals and agendas of space tourism. Before exploring fully what each of these terms means, it is important to note that a level of ambiguity is present here – some companies and individuals use these terms interchangeably, in some cases the same term means different things to different actors, and in some cases actors will use all three terms to mean simply 'space tourism' without defining nuanced difference between the three. Nevertheless, discourse within the space industry – from both the first author's research and external media sources – and the different choices of terminology combine to suggest subtle differences between the three definitions and their conceptions of what a 'space tourism' future entails.

The first of these three definitions is 'Citizen Space Exploration' which contains two combined facets. The first is an emancipatory outcome to space tourism focused around the exploration of outer space as a social space – no longer will it be only astronauts and those of significant wealth who are able to visit space, but this should rather be opened up to a range of citizens. The model of 'Citizen Space Exploration' also argues for a combination of citizen science and citizen space exploration, thereby putting forward a future where access to space has been widened, and the actors involved in space have been broadened beyond existing economic and scientific elites. As noted above, there are historical parallels with the importance of the developing aviation industries to different national narratives of citizenship throughout the twentieth century (Adey 2006b, 2007), and rhetoric around citizenship relates to wider representations of the practice of mobilities as centrally related to the expression of liberty and progress in modern societies (Cresswell 2010, 21). Although as noted above, the three definitions for 'space tourism' are sometimes used interchangeably, 'citizen space exploration' is currently the least explicitly articulated, and in some ways the most 'utopian' in its imaginings of social change, 'citizen science', and a wide access to space which is predominantly imagined as a form of living and social life, and therefore non-touristic.

The second of these definitions is 'Personal Spaceflight'. This has a strongly individualistic tone to it – the terminology implies a certain form of wealth or lifestyle akin to the assumptions one might make about 'personal trainers' or 'personal assistants' – what, therefore, does personal spaceflight seek to bring to the lives of those with the lifestyle and wealth that can afford these kinds of personal service? Would it enact the 'powerful cultural dreams of adventure' that Featherstone (2004, 2) argues are implicit in much of the discursive imaginings of automobility systems?

Would we see the alignment with existing articulations of elite mobilities (Birtchnell and Caletrío 2014), and an extension of what Edensor (2007) has named 'enclavic touristscapes', in which systems are set in place to manage and minimise the potential messiness and friction of spatial practice for the insulted travelling tourist? In the 'Personal Spaceflight' model, we see the debates on the relational experiences and affects of mobilities more generally (Adey 2006a), and the power geometries they enact (Massey 1993); it is not merely the case that some will have access to different modes of travel than others, but that the commercial and political decisions to prioritise the ease of mobility for certain parts of the population actively delimits the choices and chances of other sections of the population. In this case, we argue that the rubric of *personal spaceflight* envisages a future of *transport* – a personalised, individualistic form of transport for those who can afford it that will allow rapid travel around the world, and also leisure activities unavailable to the majority of the populace. This is in part tourism, but the tourism and leisure aspects are, in this conception, a subset of the development of a new form of transport.

The third of these definitions is simply 'space tourism' – as well as being used in the overall term for the three different understandings of the concept in this list, it is also sometimes used in a more specific sense which, like the other two, evokes a vision of a particular kind of future. Space tourism in its 'purest' form puts forward a mass-market future that emphasises the sight-seeing potential of trips into space. This is the beginning of a new form of the 'tourist gaze' (Urry and Larsen 2011) which nonetheless engages with earlier forms of tourism and a more general 'economy of signs' (Lash and Urry 1994) that recalibrate our sense of the global. Or, to extend the point, it may be that we are witnessing an iteration of the 'travel glance' (Larsen 2001), where the interactions between tourist body and passing landscape are so fleeting as to be cinematic in perception and feel. In this sense, it is a future of *commercialisation*, albeit one subtly different to existing forms of transit. Although as we shall see many in the space industry draw explicit comparisons with train travel and aviation, these are forms of travel that – aside from the initial interest when the form of transportation was new – are focused on transporting the passenger to another location. The journey itself is not the tourist attraction, unlike mass market space tourism which *does* conceive of the travel as being the attraction. This is in contrast to much of the space industry, which suffers from what Lester and Robison (2009) call the 'land bias' in space discourse – just as civilisations exploring the Earth were focused on reaching islands (and their resources) and had no interest in the water they travelled over, a 'parallel can be drawn with space' where space exploration's sole focus lay in planting a boot upon extraterrestrial soil. Space tourism, by contrast, emphasises *space*, not the physical objects it contains. This may lead to a curious absence at the heart of how space tourism is currently elaborated; the passenger, and her embodied experience of travel. In this, we have the assemblage of an actor network in which the role of the human agent is muted almost entirely and where, to draw on Larsen, tourism's 'mobile body is primarily a *moved*, rather than moving, body' ([2001, 94, emphasis in the original]). If so, then space travel differs from the imaginative definitions and affective portraits of passengers that researchers have incorporated into contemporary debates (e.g. Adey et al. 2012; Bissell 2010; Bissell, Adey, and Laurier 2011; Laurier et al. 2008). Perhaps it will be simply a matter of time before the social implications of future scenarios of space travel build more solidly on the lessons of the past, a central theme of the research findings we turn to next.

Comparisons with Existing Mobilities

Many of the study's interviewees drew explicit connections to rail travel, or (more commonly) aviation as points of comparison when trying to predict how space tourism may grow, and even as *evidence* that space tourism *will* grow based on the premise that these other forms of transportation have themselves grown to become vast industries in the decades or centuries since their inception. Visions of the future mobilities are based upon perspectives on the past and particular aspects of certain existing forms of travel and tourism.

This section will thus explore some of the comparisons and extrapolations drawn by interviewees between space tourism and established forms of mobility. Although most interviewees drew a connection between space tourism and the development of civil aviation – a comparison with an immediately appropriate logic to it – two interviewees linked space tourism with the development of rail travel instead. Whilst as we shall see the links with aviation often focused more on the development of the industry and the finance and scale involved than anything else, one of the themes in comparison with rail travel instead focused on the social changes created by forms of transportation:

> The railways enabled a lot of industries to develop because of the improved transportation – in addition to all the social changes they enabled as well – so there's probably a lot of interesting spinoffs and side effects of changes in the transportation system. [Alice, Private Sector]

This is a clear understanding of the connections between forms of transport and forms of social life. This interviewee suggested that space tourism as a form of transport could both enable new forms of industries that haven't yet been anticipated, but also make significant changes to social life. In this regard, the quote is perhaps closest to the *personal spaceflight* form of space tourism – the idea that one can offer new forms of transports that change distinct aspects of social life, such as the time to fly across the globe – but also hints towards a much wider acceptance of space tourism where existing forms of social life, leisure and tourism are changed as heavily by space tourism as they once were by rail travel (Schivelbusch 1980) or car cultures (Thrift 2004). In keeping with the comparisons with rail travel, the other interviewee instead raised an issue that was one of the two primary focuses of those who drew links towards aviation – the role of *competition* as an essential part of the growth of these other forms of transportation, and therefore a prediction that competition was essential to the success of space tourism.

> When the railways were built there was this explosion and everyone wanted to jump on the bandwagon and new railways were being built all over the place. Some were incredibly successful, some went bust. I think it will be like that. [Bob, Private Sector]

This therefore makes two closely connected but distinct predictions – firstly that space tourism as an industry will grow significantly from its current point, and that many new companies and actors will enter it, but also that these will in turn be split along stark lines of success or failure. This therefore comes with the implicit assumption that there is such a vast potential market for space tourism that a large number of actors will become involved in it (such implicit assumptions will be returned to later in this paper). However, although these comparisons to railways were drawn by two interviewees, the most common comparison was with civilian air travel. Comments about future economies, especially in terms of a fetishised and idealised representation of 'the market', were a commonality across both. However, whereas railway comparisons articulated the perceived similarities between the social impacts of early railways and the desired or predicted social effects of space tourism, the comparisons drawn with air comparisons focused on different issues. Although it seems an obvious comparison – both involve 'flight', albeit it at different heights – interviewees instead focused on themes of price and risk, rather than the theme of social impacts:

> Look at flight. When it started, it was only the top 1% flying, the domain of the mega-rich. Is that not what space is now? [...] The first aircraft that flew, single-digit passengers paying the equivalent of ten times the average person's salary was the cost of an early flight across the channel. [Carol, Private Sector]

> The situation is a bit like passenger air travel was in about 1919 when the first flights going across the English channel were a converted First World War bomber with two passengers and a load of mail, and it cost a fortune, and you might not get back in one piece [laughter]. [Dan, Public Sector]

This kind of comparison performs a great deal of discursive work. By first comparing the price ranges, the comparison argues that although the price of space tourism now is akin to the price of civil aviation a century ago, just as that cost decreased with time to the point that passenger travel is now a common occurrence, so too may space tourism. Formulated another way, this serves as a defence to a potential criticism of space tourism – it may be prohibitively expensive now, but *so was passenger air travel*. The high cost of spaceflight today, the argument goes, does not serve as evidence against the future viability of low-cost tourist space travel. Of course, as research in the mobilities literature would suggest, the idea of low-cost passenger air travel is itself subject to qualification, whether thinking about air travel as a minority privilege in absolute terms across the global population (Adey, Budd, and Hubbard 2007, 785) or within stratified Western patterns of low-budget air travel. And, we know that there are systemic inequalities that serve the mobilities of elite groups to the profound detriment of other social groups (Birtchnell and Caletrío 2014). Nonetheless, there is a teleological rendering of the space travel industry at play in the quotations here, serving to position it as *eventually* a democratically open area.

Integration with Society

There is something significant about the attempt to bring the narrative of space travel in from the remoter areas of our cultural imaginaries, and again the parallels with air travel are instructive. Adey, Budd, and Hubbard (2007, 774) 'argue that airspace is not some asocial realm or "non-place", but a space whose embodied, emotional and practised geographies remain to be adequately charted'; through the following quotes, we see the articulation of social imperatives behind the drives to normalise space technologies.

> I hope it fades to the point where it's seen as routinely useful. Rather like the motor car had a red flag at first and an aircraft was a dangerous thing to get into; it's only 100 years since an aircraft first flew. They're still high-tech and difficult to build, but they are treated as part of society. [Eve, Public Sector]

> If space is more and more important in everyday life, surely the ability to service space on a routine basis the same way as ships service ports and aircraft service passenger travel these days, it's got to be a good model for space. [Frank, Public Sector]

> The idea was to bring space transportation within the realm of normal commerce. [Grant, Private Sector]

In this way, space tourism's advocates wish it to become as normalised and as mundane as air travel, though a difference within the kind of 'mundanity' and ordinariness desired is important to note. As explored earlier in the paper, most models of space tourism conceptualise a form of tourism where the transportation and the touristic leisure experience are conflated to being one and the same. The act of boarding the space vehicle, flying into orbit, remaining there for a period of time and then returning to Earth is the experience, as well as the vehicle (both literally and metaphorically) by which that experience is delivered. Although a very small number of space tourism proposals envisage concepts far in the future of space hotels or similar, this travel-and-return model remains the leading imagining of space tourism to date. This differs in many respects from recent trends in touristic practice, especially round-the-world travel, in which an extended and autonomous mode of tourism is predicated on the capacity of the traveller to locate a sense of proximity and belonging in any part of the world, finding intimate qualities of home whilst away, as captured in Germann Molz's (2008) phrase, the 'global abode'. By contrast, these models of tourism evoke the sense of 'travelling in a mobile enclave' that Edensor (2007, 210) argues are often prompted by attempts to sanitise experience, attempts increasingly at odds with the aspirations of many tourists. Nonetheless, industry representatives hope to make the *consideration* of a space tourism trip mundane whilst ensuring that the transport–leisure conflation itself remains exciting enough to gather interest. This is a balancing act that need not be done in most other forms of tourism or transportation; the touristic option must be made to appear both exotic and ordinary, exciting but within the realm of ordinary leisure. Part of this balancing act may, according to its proponents, be achieved by upping the scale of space tourism. By shifting it away from the 'personal spaceflight' model that emphasises the *individual*, turning space tourism into a systemic phenomenon is, its advocates argue, both a route to greater acceptance and a route to a vast untapped market. One interviewee put it as follows:

> There's only roughly 50–100 flights into space per year, so the volume of traffic into space is minimal, compared to thirty million civil aircraft flights per year. The volume of traffic in aviation is orders of magnitude larger, but the potential of space we're just scratching the service. [Bob, Private Sector]

By drawing a comparison between the aviation industry – as this interviewee noted, currently totalling roughly 30 million domestic flights a year – and the nascent space tourism industry, they emphasised their belief that the space tourism industry could grow to an equivalently large size. Hype is a key part of the construction of a future technological narrative (Deuten and Rip 2000); technologies in early stages often promise total replacements of existing technologies (Brown 2003), whilst in reality they often sit alongside (Geels and Smit 2000; Giikalp 1992). Such visionary beliefs are not used for their scientific status, but rather to explicitly signal *change* (Elzinga 2004) and imply a new start or a new beginning. In this way, a prediction of scale of this sort serves to both set the ground for the scope of the anticipated project as a method of recruiting backers, but also suggests that space travel may overcome the need for aviation once it gets beyond a certain point. The improvement in travel speed remains a consistent claim across all interviewees and much of the existing space tourism literature; nevertheless, a full replacement of one form of travel with another remains unlikely.

The Importance of Competition

Another key comparison drawn with the aviation industry was the perceived importance of competition to the successful development of a space tourism future. Such a prediction that competition and deregulation are key parts of the future space industry are ones we have seen earlier in this paper (Levine 1985). Market-oriented understandings propose a simplified, deterministic and linear model of economics and technology wherein 'the market' drives innovation and technological development, a narrative which this research shows is strongly represented in the contemporary space industry. This is the model to which many space tourism advocates adhere – that developing space tourism privately, not publicly, is the optimal route to success. Outer space has been termed an 'industrial park of unfathomable size' which private actors wish to develop (Vedda 2002, 201), and as these interviewees argued, space tourism is seen as one field where this privatisation of space may become especially noteworthy:

> The essence of the project is very much in the commercial frame, to make access to space much cheaper than it is currently, and to open up the whole space market and drive it much more into the commercial domain, as with aviation. [Alice, Private Sector]

> Much more like the aircraft industry where Airbus and Boeing are slugging it out all the time, trying to get one over the other, and that way the cost of air travel is brought to a minimum because there is genuine competition. [...] So competition, genuine competition is something we think is tremendously important. [Bob, Private Sector]

The reduction of state influence in space programmes has been a key component of the growth of space tourism as it currently stands, and in turn its advocates often stress a discourse of private competition and a lack of state intervention as being crucial to the growth of the industry. In this way, not just is space tourism something that is not seen as having obvious strategic or scientific benefits for state-sponsored space programmes, but as something best suited to reach its true potential *without* state engagement, and by instead relying on the abstract concept of 'the market'. Here we might see an articulation of the logics of economic liberalisation that other researchers have identified as a characteristic of space exploration and policy (Ehrenfreund and Peter 2009; de Montluc 2009); however, this appeal to the market as guarantor of technological innovation and credibility might also serve to position space tourism as a field of the space sector that is perhaps distant from the remainder of the wider sector. When asked about whether they felt this state of affairs was likely to continue in perpetuity or whether 'the market' would be able to become the sole driver for space programmes in the future, despite the deference many interviewees displayed towards private investment in space, none were willing to predict a full marketisation of the space sector as a whole. Space tourism is thus perceived in this way as one area where public input is not required – and possibly actively discouraged – and therefore stands unique among different agendas within the space sector.

Determinism and Space Tourism

Lastly, it is important to isolate and analyse an important strand of technological determinism that remains present through most of these interviewee comments, especially in the deference to the power, influence and benefit of leaving space tourism to 'the market'. That other forms of transport experienced significant uptake and reached the point of social ubiquity is taken as evidence that a similar path can be predicted for space tourism as well; this may come down to the scale of acceptance of the form of mobility, the amount of money currently invested in it, the level of competition between different companies or providers, or a comparison drawn between the 'early days' of rail or air travel and the current early days of space tourism. These comparisons posit the ability to predict the future based upon the past (Michael 2000, 22) and that an interested observer can see clear forms of linear technological development based on this data. Some interviewees went further and described other aspects of space tourism that were perceived as equally deterministic, and in some cases the suggestion that this could be used as 'evidence' towards the future success of space tourism. One interviewee stated:

> What stops [space tourism] being cheap is simply that we don't have the technology to do that. Is the tech being developed? Yes! So logically the costs will come down. [003, Private Sector]

> Once you get to spaceplanes becoming commodities, it is the classic market economics – the more you build, the cheaper it gets, the more people you can fly, and you get this positive feedback. Once you manage to trigger that, then the industry will probably explode. [Carol, Private Sector]

This draws a commonly held connection between the costs of a technology and the time it has been developed or researched. The use of the word 'logically' suggests that it should be clear that space tourism's costs will decrease as time goes by, and that the costs and social acceptance of space tourism are dependent on the technological developments that enable it, which – likewise – will only continue to improve. Equally, the reference to 'classic market economics' performs much the same kind of discursive work – it is something *classic*, understandable and unproblematic. Once the market for space tourism has reached a particular point, this interviewee suggests, it will become self-sustaining and those backing space tourism will no longer struggle for funding or support as the competition of the market will take over, drive innovation and push prices down. Similar statements were made by another interviewee:

> We're not going to be flying around in boring old jet aircraft in 40 years time, we're going to be using this sort of technology. [Helen, Public Sector]

This returns us to Urry's analysis of the post-car future and the potential for hypermobility – in this case the promise of space tourism – or forms of sustainable localism. Both variants would agree with the statement that jet aircraft may not be fully sustainable many decades hence, but rather than postulating that new forms of localised transport will be required instead, this comment argues that new forms of technology will be implemented in their place, rather than a step 'backwards' in technological sophistication; here we have an exemplar of the 'technological fix' found in future scenarios of socio-technical life premised on models of perpetual consumerism (Urry 2011, 146). As part of this deterministic prediction, it also implicitly supports the earlier quotes that argued for an integration of space tourism and ordinary social life – if something is to replace the ordinary 'jet aircraft', it will not just be something that replaces it or improves upon it technologically, but also something that will have to take up many of the social, transportation and touristic roles that aircraft currently play.

Conclusion and Future Research Directions

This paper has presented interview data from a range of individuals with roles in a number of bodies across the space travel sector, alongside an analysis of existing space literature, examining the different models of space tourism and arguments about its feasibility that arise from previous research, as well as more generally within academic debates on aeromobilities and tourism mobilities. The paper identified three strands of anticipated space tourism, including personal spaceflight and citizen space exploration, and what each of these proposes in terms of future mobilities. 'Citizen space exploration' envisages an emancipatory form of space travel; 'personal spaceflight' appears to take the form of a more individualistic, lifestyle-oriented model, whilst 'space tourism' itself emphasises tourism, travel and a mass market appeal to the technologies and social forms involved.

The paper then explored the comparisons with existing forms of mobilities offered by interviewees. It first considered the comparisons with rail – focusing on competition and associated social changes – and with civil aviation. Interviewees suggested that space tourism was akin to aircraft roughly a century ago, and that by looking back to that period, analysts can predict with some degree of accuracy how space tourism will subsequently progress. Historical parallels with these industries in particular were emphasised in rhetorical strategies that acted as both a defence of the current costs and social exclusivity of space flight and as a way of narrating an anticipated future where space travel might be opened in a more democratic manner to wider populations. Participants emphasised the integration of space tourism with society, the perceived importance of competition and 'opening' space tourism up to the market rather than having a significant state-sponsored component, and finally the article explored the implications of many of these deterministic narratives about predicting space tourism's future(s).

However, several questions remain. The clearest uncertainty is whether there is a market for space tourism of the size that its advocates believe (or are trying to convince external actors of). One interviewee expressed an awareness of this issue, and that, despite the assumptions of space tourism advocates, there had been little research into the area done:

> There's a big assumption that space tourism is going to be attractive to people. People go on weird holidays at the best of times. Will people really want to go and spend one or two weeks in space to a holiday space station, then come back? Would they be prepared to pay for it? […] Will the average man in the street want to go spend his money on a holiday in space? [Grant, Private Sector]

The quality of the individual lived experience is thus a key question for the credibility of space tourism's proposed rise into a significant industry. As discussed earlier in this paper, space tourism futures generally imagine a form of tourism where the objective of the travel is the travel itself – to go into space, look down at the Earth from orbit, experience weightlessness – rather than reaching a distinct direction. This brings a level of emphasis onto the travel experience, and space travel comes with a number of risks (such as bone density loss) and a number of unpleasant aspects (such as G-force and nausea). Whilst many may not enjoy flying on domestic flights, this is something endured in order to reach the final destination; by contrast, the 'final destination' in a space tourism trip includes some of the downsides of the initial launch. Equally, this also entails a change in the perception of what a *holiday* or a tourist experience is – the travel and the destination become one, or at least may seem that way to the average space tourist. Questions of feasibility of space travel generally are articulated by industry representatives in terms of economic and technological issues, rather than what it might physically *feel* like to travel into space.

Ironically, perhaps, given the fantastical nature of space travel in popular culture, what the analysis reveals is a lack of imagination when it comes to thinking about the passenger; or, perhaps better, an attempt to envisage in a fleshy, embodied way who the tourists might be, why they would want to travel in space in first place and which emotions they might experience. Certain questions that point to the interrelations between affects and technologies seem to be missing. So, when out of earth's orbit, how will the space tourist's relationship with others be maintained? How will space travel intersect with the tourist's sense of belonging? Will the tourist be able to be connect with family and friends through the use of social media? What is network coverage and wi-fi like in space? These questions point to the expectations of co-presence, comfort and confidence enabled through communication technologies that tourists carry with them as they are mobile across the globe (Germann Molz 2008; Germann Molz and Paris 2015); will space tourism be a different type of tourism? There is a rich vein of research to open into the normative expectations that are implicitly driving space technologies, as so many unexamined assumptions are made about future space tourists. Future research may entail wide-ranging projects exploring whether those who are the target demographic are already significant travellers, documentary analysis of space tourism media to date, and creative methods offering ways into the imaginative visioning of eventual passengers, and their mobilities, that is co-produced with industry representatives and stakeholders.

If the question of space tourism feasibility in terms of both market capacity and physical possibility can be answered, then questions still remain about which models of space tourism will eventuate. Will we be looking at a *citizen space exploration* model, rolled out along the lines of the mass market, or will the model approximate the objectives of *personal spaceflight*, where the desires of those of sufficient wealth to fund visits beyond the atmosphere sculpt the shape of an individualised industry and, moreover, a new form of rapid transport that pulls away from those without such wealth? Germann Molz (2008, 338) has characterised many of the tourist narratives prevalent in contemporary cultures as tied to markers of privilege, such as wealth, gender, class and sexuality; will space tourism (particularly if the 'personal spaceflight' model comes to pass) offer a similar set of social dynamics? Will we observe an extension of elite forms of mobility (Birtchnell and Caletrío 2014)? In his discussion of future scenario building in transport and technological futures, Urry (2011, 141) notes the tendency by forecasters to distinguish 'between three kinds of future: possible, probable and preferable'. It is, of course, possible and perhaps probable that any eventual space tourism industry would find a path between these two extremes, but perhaps there is a question to social scientists as to whether there is a *preferable* future of space travel that we all might have a stake in envisioning. One of the overriding lessons of much mobilities research in the last decade is that mobility systems help to enact relational patterns of inequality and dynamics of power in our social worlds, not to mention harm in the physical environment; this lends urgency to debates about future mobilities that may, at first glance, appear somewhat remote or esoteric. It remains to be seen precisely what demographics may be included within space tourism; whether any eventual model of space tourism will actually resemble any one of the models identified in this paper; and the extent to which the implicit predictions about the nature of space tourists and the deterministic narratives within which they feature will come to pass. But these are social processes and debates from which social science should not be absent.

Acknowledgements

The authors wish to thank all the interviewees from the UK Space Agency and other organisations and companies who gave up their time to be interviewed for the original research. The research was not supported by funding or a grant.

References

Adey, P. 2006a. "If Mobility is Everything Then it is Nothing: Towards a Relational Politics of (Im)mobilities." *Mobilities* 1 (1): 75–94.

Adey, P. 2006b. "Airports and Air-mindedness: Spacing, Timing and Using the Liverpool Airport, 1929–1939." *Social and Cultural Geography* 7 (3): 343–363.

Adey, P. 2007. "May I Have your Attention: Airport Geographies of Spectatorship, Position, and (Im)mobility." *Environment and Planning D* 25 (3): 515–536.

Adey, P. 2008. "Architectural Geographies of the Airport Balcony: Mobility, Sensation and the Theatre of Flight." *Geografiska Annaler: Series B, Human Geography* 90 (1): 29–47.

Adey, P. 2010. *Aerial Life: Spaces, Mobilities, Affects*. Oxford: Wiley/Blackwell.

Adey, P., L. Budd, and P. Hubbard. 2007. "Flying Lessons: Exploring the Social and Cultural Geographies of Global Air Travel." *Progress in Human Geography* 31 (6): 773–791.

Adey, P., D. Bissell, D. McCormack, and P. Merriman. 2012. "Profiling the Passenger: Mobilities, Identities, Embodiments." *Cultural Geographies* 19 (2): 169–193.

Beck, U. 2002. "The Cosmopolitan Society and its Enemies." *Theory, Culture & Society* 19 (1/2): 17–44.

Birtchnell, T., and J. Caletrío, eds. 2014. *Elite Mobilities*. London: Routledge.

Bissell, D. 2010. "Passenger Mobilities: Affective Atmospheres and the Sociality of Public Transport." *Environment and Planning D* 28 (2): 270–289.

Bissell, D., P. Adey, and E. Laurier. 2011. "Introduction to the Special Issue on Geographies of the Passenger." *Journal of Transport Geography* 19 (5): 1007–1009.

Brown, N. 2003. "Hope Against Hype – Accountability in Biopasts, Presents and Futures." *Science Studies* 16 (2): 3–21.

Brown, N., B. Rappert, and A. Webster. 2000. "Introducing Contested Futures: From Looking into the Future to Look at the Future." In *A Sociology of Prospective Techno-science*, edited by N. Brown, B. Rappert, and A. Webster, 3–20. Farnham: Ashgate.

Büscher, M., J. Urry, and K. Witchger. 2010. *Mobile Methods*. London: Routledge.

Clifford, J. 1997. *Routes: Travel and Translation in the Late Twentieth Century*. Cambridge, MA: Harvard University Press.

Collis, C. 2009. "The Geostationary Orbit: A Critical Legal Geography of Space's Most Valuable Real Estate." *Sociological Review* 57 (s1): 47–65.

Cresswell, T. 2010. "Towards a Politics of Mobility." *Environment and Planning D* 28 (1): 17–31.

Cresswell, T. 2011. "Mobilities 1: Catching Up." *Progress in Human Geography* 35 (4): 550–558.

DeLyser, D. 2011. "Flying, Feminism and Mobilities – Crusading for Aviation in the 1920s." In *Geographies of Mobilities*, edited by T. Cresswell and P. Merriman, 83–98. Farnham: Ashgate.

Dennis, K., and J. Urry. 2009. *After the Car*. Cambridge: Polity.

Deuten, J., and A. Rip. 2000. "Narrative Infrastructure in Product Creation Processes." *Organization* 7 (1): 69–93.

Dickens, P. 2009. "The Cosmos as Capital's Outside." *Sociological Review* 57 (s1): 66–82.

von der Dunk, F. 2011. "Space Tourism, Private Spaceflight and the Law: Key Aspects." *Space Policy* 27 (3): 146–152.

Edensor, T. 2007. "Mundane Mobilities, Performances and Spaces of Tourism." *Social & Cultural Geography* 8 (2): 201–215.

Ehrenfreund, P., and N. Peter. 2009. "Toward a Paradigm Shift in Managing Future Global Space Exploration Endeavours." *Space Policy* 25 (4): 244–256.

Elhefnawy, N. 2004. "Viewpoint: Feeling the Pinch: Societal Slack and Space Development." *Astropolitics* 2 (3): 323–334.

Elzinga, A. 2004. "Metaphors, Models and Reification in Science and Technology Policy Discourse." *Science as Culture* 13 (1): 105–121.

Eurospace. 1994. "Small Missions: Exploiting the Opportunities." *Space Policy* 10: 322–326.

Featherstone, M. 2004. "Automobilities: An Introduction." *Theory, Culture & Society* 21 (4/5): 1–24.

Geels, F. W., and W. A. Smit. 2000. "Failed Technology Futures: Pitfalls and Lessons from a Historical Survey." *Futures* 32 (9–10): 867–885.

Germann Molz, G. 2008. "Global Abode: Home and Mobility in Narratives of Round-the-world Travel." *Space & Culture* 11 (4): 325–342.

Germann Molz, G. 2010. "Performing Global Geographies: Time, Space, Place and Pace in Narratives of Round-the-world Travel." *Tourism Geographies* 12 (3): 329–348.

Germann Molz, G., and C. Paris. 2015. "The Social Affordances of Flashpacking: Exploring the Mobility Nexus of Travel and Communication." *Mobilities* 10 (2): 173–192.

Giikalp, I. 1992. "On the Analysis of Large Technical Systems." *Science, Technology and Human Values* 17 (1): 57–78.

Hannam, K., M. Sheller, and J. Urry. 2006. "Editorial: Mobilities, Immobilities and Moorings." *Mobilities* 1 (1): 1–22.

Hannam, K., G. Butler, and C. Paris. 2014. "Developments and Key Issues in Tourism Mobilities." *Annals of Tourism Research* 44: 171–185.

Huntley, W., J. Bock, and M. Weingartner. 2010. "Planning the Unplannable: Scenarios on the Future of Space." *Space Policy* 26 (1): 25–38.

Larsen, J. 2001. "Tourism Mobilities and the Travel Glance: Experiences on the Move." *Scandinavian Journal of Hospitality & Tourism* 1 (2): 80–98.

Lash, S., and J. Urry. 1994. *Economies of Signs and Space*. London: Sage.

Launius, R. 2000. "The Historical Dimension of Space Exploration: Reflections and Possibilities." *Space Policy* 16 (1): 23–38.

Laurier, E., H. Lorimer, B. Brown, O. Jones, O. Juhlin, A. Noble, M. Perry, D. Pica, P. Sormani, L. Watts, A. Weilenmann. 2008. "Driving and 'Passengering': Notes on the Ordinary Organization of Car Travel." *Mobilities* 3 (1): 1–23.
Lester, D. F., and M. Robinson. 2009. "Visions of Exploration." *Space Policy* 25 (4): 236–243.
Levine, A. 1985. "Commercialization of Space: Policy and Administration Issues." *Public Administration Review* 45 (5): 562–569.
Massey, D. 1993. "Power-geometry and a Progressive Sense of Place." In *Mapping the Futures: Local Cultures, Global Change*, edited by J. Bird et al., 60–70. London: Routledge.
Merriman, P. 2015. "Mobilities 1: Departures." *Progress in Human Geography* 39 (1): 87–95.
Michael, M. 2000. "Futures of the Present from Performativity to Prehension" In *A Sociology of Prospective Techno-science*, edited by N. Brown, B. Rappert, and A. Webster, 21–42. Ashgate.
de Montluc, B. 2009. "The New International Political and Strategic Context for Space Policies." *Space Policy* 25 (1): 20–28.
Parker, M. 2009. "Capitalists in Space." *Sociological Review* 57 (s1): 83–97.
Parker, M., and D. Bell. 2009. "Introduction: Making Space." *Sociological Review* 57 (s1): 1–5.
Robbins, B. 1998. "Actually Existing Cosmopolitanism." In *Cosmopolitics*, edited by P. Cheah and B. Robbins, 1–19. Minneapolis: University of Minnesota Press.
Sadeh, E. 2005. "The Evolution of Access to Space as an Idea and Technology." *Astropolitics* 3 (3): 305–318.
Salomon, J.-J. 1996. "Science Policy to Cope with the Inevitable?" *Science, Technology and Society* 1 (1): 73–100.
Schivelbusch, W. 1980. *The Railway Journey: Trains and Travel in the 19th Century*. Oxford: Blackwell.
Skrbis, Z., and I. Woodward. 2013. *Cosmopolitanism: Uses of the Idea*. London: Sage.
Thrift, N. 2004. "Driving in the City." *Theory, Culture & Society* 21 (4/5): 41–59.
Urry, J. 1990. *The Tourist Gaze*. London: Sage.
Urry, J. 2000. *Sociology Beyond Societies: Mobilities for the Twenty-first Century*. London: Routledge.
Urry, J. 2007. *Mobilities*. Cambridge: Polity.
Urry, J. 2011. *Climate Change and Society*. Cambridge: Polity.
Urry, J., and J. Larsen. 2011. *The Tourist Gaze 3.0*. 3rd ed. London: Sage.
Vedda, J. 2002. "Space Commerce." In *Space Politics and Policy*, edited by E. Sadeh, 201–229. London: Kluwer.
Williams, R. 2006. "Compressed Foresight and Narrative Bias: Pitfalls in Assessing High Technology Futures." *Science as Culture* 15 (4): 327–348.

Accelerate, Reverse, or Find the Off Ramp? Future Automobility in the Fragmented American Imagination

KATHERINE G. REESE

School of International Service, American University, Washington DC, USA

ABSTRACT *How is the future of automobility envisioned in US policy discourse? This paper examines the policy documents of the US Energy Department, Transportation Department, and Environmental Protection Agency, as well as the writings of the post-carbon movement. Using discourse analysis, the paper investigates how, in narrating the future of the automobile in the US, these texts perform political work: producing forms of subjecthood and legitimating action. It concludes that the dominance of automobility in the American imagination is being unsettled as discourse about the future of the automobile fragments into three distinct narratives of progress, return, and radical change.*

What is the future of the car? Will we create a high-tech world for ourselves, with low-carbon vehicles plugged into a renewable energy system? Will we rebuild cities to recover forms of neighborhood lost when the automobile came to dominate the landscape? Will we find car use inevitably and drastically curtailed on a post-peak-oil planet?[1]

These questions are not merely about the technological future of the automobile. As the mobilities field has so clearly established over the course of the past decade, mobility cannot be simply equated with transportation technologies. How we move through the world has profound constitutive effects on social life and cultural meaning. The literature on automobility in particular has illuminated how the automobile has helped to create the landscape of the modern West and the subjects who move through that landscape. The future of the car, then, means the very future of the modern West.

In the past several years, as global climate change and energy instability have claimed ever more urgent public attention, automobility literature has turned a focus toward the future. Thus far, scholars have sketched out potential transformations of automobility, bringing a degree of speculation to empirical evidence of already-changing trends. This paper, however, takes a different approach to the question 'what is the future of the car?' Rather than answering the question, it analyzes the act of asking. Here, the 'future' is not an eventuality to be anticipated, but a discursive tool to discipline or mobilize action in the present. This paper sheds light on the political moves that actors make when talking about the future of the car – moves that often have less to do with the automobile than with expert authority and social identity writ large.

As the first half of the paper argues, envisioning a future necessarily means making assumptions about what can change and what will continue; as such, envisioning a future legitimates certain forms of subjecthood, power, and action. This is particularly the case in 'authoritative representations' – texts such as design standards or technological 'roadmaps' that officially define what will be possible and desirable in the future. In the second half of the paper, I identify three narratives in US policy discourse that envision a future transformation of the automobile. One reproduces the dominant modern narrative of progress. It encourages society to 'accelerate' – to increase the speed of scientific research and technological development. The second taps into long-standing critiques of modern alienation and acceleration. It sees the need to 'reverse' the excesses of late twentieth-century automobile-dominated planning and sprawl. And the third draws on a dystopian environmental tradition. It posits that communities should 'find the off ramp,' adapting to life in a dramatically different post-peak-oil future before it is too late.[2]

Landscapes are shaped as much through discursive legitimation as through the pouring of asphalt. Narratives of the future matter because they have political effects. Besides justifying the production and use of different forms of mobility, they articulate different forms of subjecthood, from consumers to urban dwellers to community activists. They legitimate different configurations of expertise and state authority. Moreover, these narratives matter because taken together, they point to a deeply fragmented American imagination. Americans are not merely disagreeing over, for instance, when electric car sales might increase or whether bike lanes should expand. They are profoundly conflicted over who they are, both collectively and as subjects moving through the world.

Situating Automobility in the Twenty-first Century

Automobility in Modernity

In the years since the concept of 'automobility' came to encompass the broader political and sociocultural dimensions of automobile production and use (Sheller and Urry 2000; Urry 2004), it has served to expand the scope of scholarly analysis of the car. This scholarship has produced the understanding that automobility has been both 'product and producer of modernity' (Rajan 2006, 113), playing a central role in accelerating and disembedding life in the modern era (Beckmann 2001; Urry 2004; Paterson 2007; Seiler 2008). Fueled by energy-intensive economic expansionism and justified by the claim that modern societies need good roads (see Urry 2013 on 'carbon capital'), automobility changed the Western urban landscape. Car use allowed for the (further) rationalization, segregation, and specialization of urban space (Featherstone 2004; Bonham 2006). The building of highways contributed to the construction of national-scale geographies, particularly in Germany and the US (Goddard 1994; Koshar 2004). Automobility influenced modern environmental imaginaries of wilderness and human settlement (Jackson 1985; Flink 1988; Sachs 1992; Gunster 2004; Huijbens and Benediktsson 2007; Clarsen and Veracini 2012), while contributing to the proliferation of the 'non-places of super-modernity' (Merriman 2004; Urry 2004, 30; cf. Wollen and Kerr 2002 or Dalakoglou and Harvey 2012).

By the end of the twentieth century, in many Western nations, there were half as many cars as people (World Bank 2003, 167), while the average urban dweller drove thousands of kilometers annually (Kenworthy and Laube 1999).[3] Automobility now touches the everyday habitus of millions of people, as the textures and tempos of their existence are constituted by moving through an auto-dominated landscape (Bull 2004; Bonham 2006; Hagman 2006; Urry 2006; McLaren and Parusel 2012). Whether one is a regular driver or not, automobility – as an emotional, ethical, embodied experience – shapes how one experiences the world (Dant 2004; Sheller 2004; Pearce 2012). The car and its rituals serve to perform group belonging and provide outlets for self-expression (Miller 2001; Gartman 2004; Collin-Lange 2013; Lumsden 2013). Driving embodies that quintessentially modern subject: the individual. Automobility is 'the (literally) concrete articulation of liberal society's promise to its citizens that they can freely exercise certain everyday choices' (Rajan 2006, 113–114; see also Latimer and Munro 2006). Yet, as Paterson (2007) and others have shown us, the modern state has been deeply implicated in the production of this individualized, autonomously mobile subject (Bonham 2006; Merriman 2006; Packer 2008; Seiler 2008).

Automobility in Crisis

Altogether, the literature has established that automobility has given form to the modern landscape and the subjects moving through that landscape. Yet the literature has also shown automobility to be deeply problematic as it 'epitomizes both the "goods" and the "bads" of modernity' (Hagman 2006, 64). It is riddled with 'impossibilities' (Bohm et al. 2006), 'discontents' (Paterson 2007), and 'broken promises' (Hagman 2006). While different scholars place different emphases on the 'automobility problematique,' to use Cohen's (2006) term, a core litany of concerns is often evoked: congestion, violence against humans and non-humans, social isolation, dependence on oil, extensive land use, and global climate change. While the automobile knows a long history of intermittent crisis and critique, coming under fire for its social and environmental impacts (Mumford 1961; Nader 1965; Freund and Martin 1993; Kunstler 1994), the past decade's critical scholarly interest in automobility 'suggests that cars and driving, until recently rendered mundane by their ubiquity, constitute once more something of a crisis' (Seiler 2010, 389).[4] We have 'entered a phase of doubt' (Flonneau 2010, 384), where scholars aim to 'expose and oppose' (Bohm et al. 2006, 11) automobility as they intentionally 'trouble the car' (Conley and McLaren 2009).

However, I would underline the specifically *temporal* inflection of critique in the past decade. The literature has moved from mere critique of the automobile and its effects to the claim that *automobility cannot continue*. Bohm et al. (2006) call this the 'impossibility' of automobility: 'if continued, a car-based regime generates widespread problems ... which cannot be resolved without abandoning the regime itself. In this sense *the continuation of automobility is impossible* in its current form' (Bohm et al. 2006, 9; my emphasis). Similarly, Beckmann argues that automobility 'has caused incalculable risks on a global scale and is subsequently *threatening its own foundations*'; it 'has turned against itself' (Beckmann 2001, 604; my emphasis; see also Beckmann 2004). Concerns about automobility necessarily become concerns about the future.

As such, scholars have begun to ask what the future of automobility can and should hold. Some extrapolate economic and demographic trends to speculate about 'peak car' (Goodwin and Van Dender 2013; Metz 2014). Several focus on possible transformations of automobility as a socio-technical system (e.g. Geels 2012), with the most comprehensive work coming in Dennis and Urry's (2009) *After the Car*. Others, placing more emphasis on human agency, have assessed the potential for shifts in the social and cultural practices that constitute automobility (Sheller 2012; Watson 2012; Dowling and Simpson 2013; Kent and Dowling 2013). All these studies attempt to analyze 'openings' or cracks in the automobility regime (Sheller 2012, 186) and speculate on those openings' expansion into new systems and new material cultures. They share the general consensus that though automobility may be increasingly destabilized (particularly in urban areas), the future remains uncertain.

Such scholarship on the future of automobility is commendable and necessary. Dennis and Urry are quite right when they state that, given the threat of global climate change, '[t]hinking the future' of automobility 'simply cannot be avoided' (Dennis and Urry 2009, 147). However, there is another, equally useful way for scholars to engage with the future of automobility. Here, I would invite mobilities scholars to treat automobility as 'a story,' in Cotten Seiler's words, 'that a culture tells itself about itself' (Seiler 2010, 392). From this perspective, it is no more meaningful to ask what tipping point might produce what kind of post-car world than it is to ask what stories our cultures are telling about themselves. The goal of such scholarship is not to envision the future of automobility, but to analyze the political discursive work that goes on when actors envision the future of automobility.

The Politics of the Future

Despite the centrality of the ideas of time and space in the mobilities literature, work thus far has tended to privilege space – analyzing how movement demarcates and gives meaning to space (see Jensen 2011). Yet time – specifically *historical* time, as opposed to personal time – is equally bound up in the production and experience of mobility. All forms of mobility are richly imbued with a broader social narrative, accessing 'structured stories that explain' – and, I would argue, justify – 'the existing material culture' (Sheller 2012, 185). To move through the modern world is to tell a story about where we are all headed. For much of the twentieth century, this was a story about progress and order. State authority governed mobilities spatially (Urry 2007; Jensen 2011; Bærenholdt 2013) – and this governance was justified by the state's claim to be able to bring some degree of order and predictability to the future (MacIntyre 1984; Reese 2015).

Just as the self is constructed in terms of the other, or mobility is conceived of in terms of moorings (Adey 2006), the future is imagined through a dual logic of change and continuity. Envisioning a future implies change: it allows one to problematize aspects of the self and to imagine transformation. This ability to envision potential change is at the heart of creativity and critique; indeed, it is a core aspect of human agency itself (Ricoeur 1991; Joas 1996; Emirbayer and Mische 1998). Yet one can only imagine change by also expecting continuity. Envisioning a transformation to the urban built environment requires one to expect cities still to exist in the future. One can only imagine a 'car of the future' if one expects drivers still to exist. It sounds obvious – and that is precisely the point. To narrate such continuities is to define the boundaries across which future transformation cannot extend. Expecting future continuities thus often plays upon (and reproduces) taken-for-granted structures of social and political life.

Clearly, we can expect both continuities and changes in the future – but there is a more fundamental ontological argument here. We envision the future by interpreting certain aspects of the collective self as (functionally) permanent and others as changeable. When we do so, we reproduce *in the present* certain forms of subjecthood (along with the power relations in which they are enmeshed). Narrating a better future gives meaning to action now, legitimating some forms of action and characterizing others as irrelevant or counterproductive. From this perspective, it is not that what we do today shapes what tomorrow will look like. Instead, what we think tomorrow could look like shapes what we do today.

In the textual analysis that follows, I focus on discursive representations of the future in US policy discourse. Specifically, I analyze how discursive representations of the future invite people to be or become particular kinds of subjects in the present. By delineating particular social roles – complete with expected behaviors and associated meanings – these representations of the future can define the range of possible subjects we can be. When actors discursively articulate social roles, they 'define subjects authorized to speak and to act' (Milliken 1999, 229). In the context of the future of automobility, this means that some types of actors (such as highly trained engineers or well-pedigreed urban planners) are represented as uniquely authoritative, in a position to know and shape the future. Discursive representations of the future also (implicitly or explicitly) incorporate these subjectivities into larger story arcs; they construct a potential world in terms of intelligible beginnings and endings, causes and effects, problems and solutions.[5] My textual analysis thus focuses on subjectivities and narratives, with particular attention paid to representations of change (who is seen as capable as effecting change? What is a legitimate object of change?) and continuity (what is seen as unchanging from the present to the future?).

Narratives in Authoritative Texts

Current policy discourse in the United States narrates the future transformation of automobility in three specific ways, which I characterize here as 'accelerate,' 'reverse,' and 'find the off ramp.' The first envisions accelerating technological changes to the automobile itself; the second sees a return to the built environment of the pre-automobile era; and the third hopes for a new trajectory for community in a future defined by peak oil and climate change.

The rest of this paper explores how these three different narratives are inscribed in three sets of texts.[6] The first corpus of texts, those that articulate the 'accelerate' narrative, is built around technological 'roadmaps' produced by and for the advanced vehicle technology programs in the US Department of Energy (DOE).[7] The second corpus of texts, articulating the 'reverse' narrative, is built around documents codifying 'smart growth,' produced by the US Department of Transportation (DOT) and Environmental Protection Agency (EPA) as well as their civil society partners.[8] I built these two particular sets of texts because, while there are many modes of envisioning the future (from apocalyptic fiction to actuarial tables), the relationship between state power and discourses about the future is uniquely clear in 'authoritative representations,' those particular characterizations of the world taken to be official, formal, or standard (see Jasanoff 2004 on Scott 1998 and Anderson 1991). In these texts, the state performs itself as an authority and thus reaffirms its claim to order the future for the benefit of the public: knowledge production and governance become bound up with representing what is (and will be) possible and desirable.

In contrast, the third corpus of texts analyzed here – a set of documents which urge society to 'find the off ramp,' or transition now to a post-carbon future – is comprised of texts that are not meant to be authoritative. They fall outside the position of state-sanctioned expertise. This is in part because the third narrative does not have the same (national-level) institutional presence as the other two narratives. Besides serving as a foil to the more authoritative state representations of the future of automobility, this narrative is included because it is gaining traction in civil society and among policymakers at the local level. Its organizational home in the US is Transition US, the US headquarters of the broader Transition Movement.[9] The third set of texts, then, is made up of writings published or recommended by Transition US.[10]

Each of these three sets of texts, from varying positions of authority, narrates the future of automobility differently. As I will elaborate in the conclusion, advanced vehicle research, smart growth, and Transition are not options to choose among, nor horses to bet on. They are three sites where the US tells the story of its collective self.

Part 2. Automobility in the Twenty-first-century American Imagination

Accelerate

The first corpus of texts is made up of roadmaps written to provide clarity on prospects for advances in vehicle technology. Reflecting the dominant modern understanding of unidirectional progress, these texts envision a future populated by autonomously mobile drivers, automotive engineers, and renewable energy technologies, prominent among which are light and efficient cars that run on renewable energy. In this 'Promethean' (Dryzek 2013) view, history is an upward trajectory of scientific research and technological development, where each passing year brings humanity closer to a better future. In fact, in these texts, research will not only make a better future, but must make it better *as quickly as possible*. Like the concept of acceleration itself, the vision here is not merely of progress, but of progress in how fast society progresses.

These envisioned changes are directed towards managing two 'challenges,' in the DOE's terms. Evoking threats of price volatility, political instability in oil-producing regions, and increased demand in the developing world, the texts situate the US within a context of a limited and unreliable oil supply (US DOE 2006, 2010, 2011). Yet, these texts suggest that the limits of oil supply are negotiable; human innovation can find a way to make the most out of what there is. The second 'challenge' that orients the texts is global climate change, which they see having potentially large-scale long-term consequences (US DOE 2011, 2006; America's Energy Future

Energy Efficiency Technologies Subcommittee, National Academy of Sciences, National Academy of Engineering, and National Research Council [AEF Panel et al.] 2010), though these consequences can be predicted, quantified, and managed (US DOE 2011). These challenges imbue the narrative with a sense of urgency; they set 'a relentless clock on our actions,' in the words of a former Undersecretary of Energy (Koonin 2011). This means that the goal of advanced vehicle research becomes not merely to drive 'evolutionary' change in technologies, but to accelerate the rate of change through an acceleration of scientific research itself (US DOE 2010; National Research Council [NRC] 2013; US DOE 2011). In this way, the texts deploy the most powerful tool in the modernist conceptual toolkit: accelerated progress through more knowledge production.

The texts envision change primarily in technological terms. They anticipate small-scale short-term changes: efficiency gains in the internal combustion engine and increased use of hybrid vehicles (Oak Ridge National Laboratory 2010; AEF Panel et al. 2010; US DOE 2011, 2012). These small-scale changes are seen as appropriate precisely because they represent 'evolutionary' and not radical transformations of the energy system (US DOE 2011, x; see also US DOE 2006, 2010; NRC 2013; AEF Panel et al. 2010; DOE 2009). The texts do expect more drastic transformations in the long term: adoption of plug-in electric vehicles and, eventually, hydrogen fuel cells (e.g. AEF Panel et al. 2010; US DOE 2006, 2012; NRC 2013; US DOE 2011). Moreover, these changes are envisioned as part of a broader effort to 'accelerate the large-scale transformation of energy production, delivery, and use to a low-carbon energy system' (President's Council of Advisors on Science and Technology [PCAST] 2010; vii; see also AEF Panel et al. 2010, vi). The texts thus envision a long-term transformation of the entire US energy landscape.

Yet, this transformation is envisioned in terms of three important continuities. The first is that of the driver as a (consuming) subject. There is no suggestion that cars might someday no longer play a central role in the economy and daily life; in fact, these texts expect demand for automobiles to remain steady in the US while increasing globally. The driver-consumer plays a pivotal role in the envisioned technological change: producing an advanced vehicle will only solve the challenges of oil consumption and climate change if a significant number of consumers buy and use that advanced vehicle (US DOE 2011; 32; US DOE 2010; AEF Panel et al. 2010). Yet driver-consumers are treated as fixed, risk-averse, and unlikely to embrace change unless it is in their economic interest (US DOE 2010; AEF Panel et al. 2010). They are understood to be drivers, and American drivers at that: they like their cars big, comfortable, and fast; and they will continue to do so well into the future (e.g. NRC 2013). This expectation sets the bounds for possible transformations of the automobile (for instance, it precludes the possibility of much smaller vehicles; US DOE 2011; AEF Panel et al. 2010). The American driver-consumer will not change, so technology must (US DOE 2010, 2011; AEF Panel et al. 2010).

This is where the second continuity comes in: the automotive engineer. Because the driver-consumer's behaviors and preferences are ostensibly predictable, the engineer can design around those behaviors. The engineer is thus the one who makes it possible for technology to reconcile the driver-consumer's (oil-consuming and carbon-producing) behavior with the goals of reducing oil consumption and carbon emissions. The engineer is the agent of change. While the driver-consumer is not expected to learn or change, the engineer experiments, discovers, and innovates in a high-tech state-funded laboratory, bolstered by many years of training. Moreover, the engineer in question is specifically and only an *automotive* engineer: change is not directed at anything other than the automobile. Though some advanced vehicle texts refer to other modes of transportation, they do so only to immediately abdicate responsibility for such alternatives (US DOE 2011; AEF Panel et al. 2010; US DOE 2010). In doing so, they functionally equate knowledge production exclusively with changes to the automobile; they eliminate space for considering technological developments in broader mobility terms.

The third continuity expected in the texts is the nation-state as the locus of problem-solving. The role of the state here is not to coerce, but to facilitate knowledge production. In this narrative, addressing oil consumption and climate change necessarily requires research on a large scale; it requires advanced technology to create a vehicle that reduces oil consumption and carbon emissions while still meeting the expectations of the American driver-consumer (e.g. US DOE 2012). As such, only the nation-state has the resources and authority to produce, centralize, and order knowledge production (by identifying targets for technological development or setting research priorities [e.g. Oak Ridge National Laboratory 2010; US DOE 2011]) on the necessary scale.

Altogether, the project of the advanced vehicle envisions accelerated technological progress towards a transformed energy landscape. As these texts narrate this progress, they reaffirm the boundaries of a particularly capitalist, particularly autonomous and mobile American identity. The driver-consumer is sovereign over her or his vehicle choices, choices whose impact the highly trained engineer minimizes through scientific innovation. These texts justify centralized knowledge and state wealth, implying that such forms of power are necessary to effect the kind of large-scale high-tech change needed to reduce oil consumption and carbon emissions. The state thus shores up its power by way of a particular concept of the future – namely, a future where only the state can effectively marshal human ingenuity (via the expert) to secure a better future.[11]

Reverse

The second narrative of the future of automobility de-emphasizes the car itself and envisions a built environment that accommodates bicycles, walking, and public transport options. This corpus of texts is drawn from 'smart growth' efforts within US government agencies and their civil society partners.[12] While advanced vehicle research focuses on technological changes to the automobile itself, smart growth seeks to reduce the dominance of the automobile in the built environment through urban design emphasizing mixed land use, density, walkability, and a wide range of transportation choices.[13] Though smart growth embraces the future in many ways (e.g. planning for growth), it often narrates the future in terms of the past. In this narrative, the post-war era of suburban expansion and motorization was a moment of rupture in the built environment – and therefore the social fabric (e.g. Smart Growth Network, International City/County Management Association, and US Environmental Protection Agency [SGN et al.] 2002, 25). The future, in this case, represents the possibility to 'reverse' – to rebuild the landscape (and society) that was lost with the rise of the automobile. As such, the smart growth narrative makes a somewhat paradoxical claim both to progress (a better future) and to return (a future that looks like the past). In this, the narrative draws on a long tradition of resistance to modern industrialization and alienation, from the nineteenth-century Romantic movement through William

Morris, Ebenezer Howard, and more recent deep ecological thought.[14] In this tradition, a better future is secured not through ever-accelerating technological innovation, but by conserving and rehabilitating continuity in the landscape.

While the 'accelerate' narrative is oriented towards issues of oil consumption and climate change, this 'reverse' narrative sees a future where smart growth solves a wide range of automobile-fueled problems (American Planning Association [APA] 2006; SGN et al. 2002, i) and does so with a degree of permanence (Federal Transit Administration [FTA] 1999). Smart growth introduces alternative modes of transportation that reduce driving and therefore greenhouse gas emissions (US Green Building Council [USGBC] 2009; Duany, Speck, and Lydon 2010). It includes 'traffic calming' design elements (such as narrow lanes or trees) that slow traffic and make the roads safer (US Environmental Protection Agency [US EPA] 2009), and elements (such as inviting public spaces) that 'support stronger communities through improved social interaction' (SGN et al. 2002, 26). In the texts, these transformations are imbued with a sense of return. The envisioned change means '*re*integrating the components of modern life ... into compact, pedestrian-friendly, mixed-use neighborhoods' (APA 2006, 602; my emphasis) and '*[r]e*storing the centrality of the neighborhood' (Duany, Speck, and Lydon 2010; xvii; my emphasis). It means recreating 'Main Streets' and engaging in Traditional Neighborhood Development, 'a style of development that works to emulate many of the features of urban neighborhoods of 50–100 years ago' (APA 2006; 602). Yet, this is a return not to a specific time, but to an ostensibly timeless way of doing things, as the walkable neighborhood 'has been the fundamental increment of human settlement throughout history, interrupted only by the 60-year aberration that we now call suburban sprawl' (Duany, Speck, and Lydon 2010, 1.3). The USGBC (2009) quotes Lewis Mumford to claim that 'neighborhoods, in some primitive, inchoate fashion exist wherever human beings congregate.' Smart growth thus means neighborhoods not (just) as they once were, but as they (supposedly) have always been.

The texts also envision changes to the practice of urban planning itself. Planning in the mid- to late-twentieth century meant creating wholly new, futuristic developments on a *tabula rasa* – building roads and suburban subdivisions on greenfields (or in urban areas cleared by razing low-income neighborhoods). In contrast, smart growth sees itself rehabilitating or conserving the existing built environment, rather than creating something entirely new (see USGBC 2009). The kinds of projects that are lauded are those that actively engage the past, reintroducing lost elements (e.g. streetcars) and working with inherited constraints (e.g. empty lots or abandoned elevated train lines). When the Smart Growth Network writes about innovation, it does not mean cutting-edge technologies; it means new legal tools to protect wildlife habitats and new zoning codes that make it easier to build on brownfields (previously developed and/or potentially polluted land) than on greenfields (SGN et al. 2002). Innovation means finding new ways to recreate what was lost and to keep intact what still exists.

This 'reverse' narrative envisions change in terms of three continuities. The first of these is the neighborhood inhabitant. Like the driver-consumer in the section above, the neighborhood inhabitant is mobile; however, for the inhabitant, mobility is not precisely autonomous. Rather, it is shaped at least in part by the landscape, reliant upon the infrastructure that the urban planner designs and the state builds. If the inhabitant uses an automobile, this use is influenced by the conveniences afforded by the built environment (such as extensive free parking), rather than ontologically prior individualistic preferences. If the built environment affords it, the inhabitant is sometimes a passenger, sometimes a cyclist, sometimes a pedestrian (e.g. Duany, Speck, and Lydon 2010; US EPA 2009; American Association of State Highway and Transportation Officials [AASHTO] 2004). This plurality of mobilities allows the inhabitant to interact with others and to feel connected to the neighborhood (SGN et al. 2002; USGBC 2009; APA 2006; Duany, Speck, and Lydon 2010). These texts thus narrate a future where smart growth produces lower carbon emissions and less oil consumption, but also results in people having a sense of place. The texts thus challenge the long-dominant American idea that, in order to be happy, one has to be able to escape to one's own suburban domain (e.g. US EPA 2009).

The neighborhood inhabitant is different from the driver-consumer in a number of ways, yet the two are similar in one crucial aspect: both are seen as predictable. For instance, the inhabitant will never want to walk more than half a mile to reach transit (APA 2006, 450; USGBC 2009, xvi; SGN et al. 2002, 10; Duany, Speck, and Lydon 2010; AASHTO 2004, 96), and will always be more likely to walk if the sidewalk is pleasant, well lit, direct, and does not require the use of an overpass or underpass (SGN et al. 2002, 31; AASHTO 2004, 96; Duany, Speck, and Lydon 2010, 3.5; APA 2006, 279). The behaviors of inhabitants are patterned, and those patterns can be known and designed for. This is where the second continuity, the urban planner, enters the scene. Urban planning is premised on design as a structuring of possibilities – design as the skillful manipulation of people's impulses towards beauty and convenience in the service of socially desirable behaviors. Though many of the texts acknowledge the importance of public input on design (APA 2006, 46, 55–6; SGN et al. 2002, ch. 10), it is the urban planner who views the landscape critically, envisions change, 'educates' the public about the need for change (SGN et al. 2002 10, 82; Duany, Speck, and Lydon 2010, 1.3), and ultimately acts as the agent for transformation. 'Livable communities do not simply happen,' as the FTA tells us, but require the intervention of expert authority. In this, the urban planner mirrors the role of the automotive engineer above: both are engaged in designing a better future. Both design around expectations about who people fundamentally are and will be, and how they would move through the world if given the opportunity. Yet the planner exerts power with a more Foucauldian touch. While the automotive engineer seeks not to guide the driver's behaviors but to reduce the impact of those behaviors, the smart growth planner designs the neighborhood to encourage better (lower-carbon, safer, less-consuming) behavior from its inhabitants. Design functions as governance; like prison design or highway design, neighborhood design canalizes movement to produce a specific social outcome.

Lastly, the neighborhood inhabitant and the urban planner are situated in a context of state authority. In contrast with the nation-state discussed above, the state here is significantly more diffuse and multi-layered (not least because, by its nature, transportation and land use policy are issues for local government). Yet the state in these texts positions itself similarly: as a centralizer and legitimizer of knowledge. The knowledge here takes a slightly different form – instead of research findings about lithium ion batteries, the state is dealing in standards for transit headways and curb heights. Yet there is a parallel logic of standardizing, universalizing, and re-diffusing knowledge in order to scale up change (US EPA 2009; US Department of Transportation [US DOT] 2010). This leads to a certain irony: the ideal smart growth neighborhood is a small, tightly knit, unique neighborhood with a sense of history and place; yet building such neighborhoods effectively and widely means, in these texts, creating a national, state-legitimated standard for such neighborhoods. Though the urban planner is the one to effect change, the state (due to its resources and authority) is the one to secure the conditions for this change at a meaningful scale.

Find the Off Ramp

The third narrative envisions a world radically and inevitably different. In these 'post-carbon' texts, the way of life that many have come to take for granted in the industrialized West – automobility included – will end whether we want it to or not. In this narrative, automobility, like all fossil-fueled aspects of life in the early twenty-first century, will change dramatically as the era of cheap energy ends and climate change worsens. In these texts, the climate is already changing (Transition United States [Transition US] 2012) and peak oil is imminent (Heinberg 2003; Hopkins 2008). Collapse will be sudden and large scale: 'we are facing a systems failure' (Transition US 2011, 9; see Heinberg 2003). The change envisioned here differs from the other two narratives, partly because it is considered potentially catastrophic in scale (Heinberg 2003; Hopkins 2008) – and partly because it is considered inevitable, unpredictable, and out of the ability of humans to control. Despite claiming to avoid apocalyptic tropes, these texts tap into a dystopian strand of environmental thought, one that warns of imminent ecological catastrophe leading to a radically different (but potentially better) world.[15] These texts narrate a future that may well be grim, but they temper this with optimism about the ability of humans to adapt in a 'relocalized' post-carbon world.

The first change envisioned by the texts is 'energy descent,' a reduction in the amount of energy used in the industrialized West (Heinberg 2003; Hopkins 2008). This reduction is considered inevitable because it is driven not by human action but by geological and ecological forces, and those (presumably) cannot be ignored (Heinberg 2003; Chamberlin 2009). As fuel becomes scarce, according to this narrative, social and economic life will relocalize (i.e. contract geographically; Hopkins 2008; Gilbert and Perl 2010). This means a significant drawing down of automobility. Resources for building automobiles and roads will become less reliable and more expensive, and burning fossil fuels may become socially unacceptable in an era of runaway climate change. Whereas the smart growth texts see the dominance of the automobile in the built environment gradually designed away, the post-carbon texts see it simply falling apart.

Though nothing can be done to avoid the energy descent, these texts argue, much can be done to prepare for it – and that means intentional 'transition.' As Transition US (2011) tells us, '[c]hange is happening–our choice is between a future we want and one which happens to us.' This intentional transformation of society is envisioned to be large, systemic change – 'extraordinary and historic' (Transition US 2011, 21). The decisive break afforded by a rapid energy descent means a change so radical that no twentieth-century expertise can claim what the future can and should hold; attempting to plan a post-carbon future is akin to 'trying to create a 'parallel universe'' (Transition US n.d.). The radically different nature of a post-carbon, post-peak-oil, nearly post-apocalyptic world opens up the scope of potential human effort; these texts thus see space for human agency even in a collapsing world.

For automobility, transitioning to a post-carbon world means making a number of transportation changes now. Interestingly, many of the post-carbon policy proposals would not look out of place in a smart growth document. They include elements such as prioritizing bike lanes; providing more bike parking; adopting a bike-sharing program; decreasing free parking; increasing pedestrian-friendly spaces; encouraging car-sharing; and expanding and improving transit (Portland Peak Oil Task Force 2007; Berkeley Oil Independence Task Force 2009; Bloomington Peak Oil Task Force 2009; San Francisco Peak Oil Preparedness Task Force 2009). The texts recommend building dense, walkable, transit-oriented development. However, the difference between these texts and smart growth texts comes in the narratives that orient these policy recommendations. In these post-carbon texts, the changes to the built environment are not to recapture a pre-automobile past but to survive in an inevitably post-peak-oil world. In such a world, governments may not be able to afford to maintain road infrastructure (Portland Peak Oil Task Force 2007, 39). 'When gasoline prices sharply rise and supplies become unreliable,' the San Francisco energy descent plan states, a change in transportation patterns 'will occur regardless of whether San Francisco plans for it or not' (63). This narrative imbues change with an urgency that smart growth discourse lacks: '[*t*]*here is no time to lose*' (San Francisco Peak Oil Preparedness Task Force 2009, 5; original emphasis).

Even when envisioning a near-apocalyptic future, these texts suggest that one thing will not change: the capacity of humans and communities for self-transformation. The first form of subjecthood that the texts evoke, then, is that of the adaptable human: a fundamentally creative creature who learns, experiments, and thrives in hard times. Interpreted in a slightly more grim fashion: the only human to survive in a radically different future will be the one who adapts. In this view, one may be a driver now – one may have enjoyed automobility along with all the other conveniences afforded by modernity. Yet if one is no longer a driver in a post-carbon future, it will not fundamentally change one's identity. One can find other ways to move through and be in the world. If the joys of a road trip is no longer available, one may yet learn to 'appreciate the quality gained by slower motion' such as on a train trip (Gilbert and Perl 2010, 9). One may learn to appreciate the benefits of stillness, feeling 'more connected to [the] immediate area, more intimately acquainted with its nooks and crannies' (Hopkins 2008, 113).

The texts also envision continuity in the form of the resilient community. Resilience is understood here effectively to mean long-term self-sufficiency, a community's 'ability to function indefinitely and to live within its limits' (Hopkins 2008; 13). In a world where the global systems that now sustain everyday life (oil production and distribution, manufacturing and shipping) disintegrate, the only community that continues is one that disengages from global commodity chains. It produces much of its own food and builds its houses from locally sourced materials. It knows its own ecological assets, its water sources, and its native species, and it fosters those assets so as to avoid relying upon distant natural resources. In this view, the resilient community is not necessarily immobile – people still cycle, share cars, and ride in buses. But they know, for instance, how to fix their own bikes (e.g. Transition US 2011, 11) and produce their own biodiesel (e.g. Hopkins 2011, 56). Their mobilities do not rely on resources or expertise from outside the community.

In contrast to the texts discussed in earlier sections, these texts express uncertainty about the nation-state's capacity in the future. Several suggest that state resources simply will not be available in the future – and may already be declining (Heinberg 2003; Bloomington Peak Oil Task Force 2009). They claim that national-level governments cannot (or will not) avert the worst effects of climate change and peak oil

(Transition US 2011). In such a future, local-level communities have to act because the state will not. As such, the locus of problem-solving is not the nation-state but the community. While these texts do rely heavily on expert authority to make the case for imminent collapse, they break from the authority of science and the state when it comes to what to do about imminent collapse. Having characterized humans as fundamentally adaptable creatures, the texts are able to make an unusual but emancipatory discursive move: they claim that expertise has no privileged knowledge of what can happen in a radically different future. Past behaviors are no guide to future potentialities, and so the space of possibility becomes wide open.

Conclusion

There are three narratives at work in US policy discourse in the twenty-first century. One sees a future where the automobile's contributions to climate change and oil depletion are dealt with by an accelerating program of technological automotive research. The second narrative suggests that the best approach to the social ills unleashed by widespread automobile use is to 'reverse,' to return to the social and built landscapes of the (imagined) pre-automobile era. Though the historical arcs of these narratives differ, each envisions change effected by experts embedded in continued structures of state authority. In contrast, the third narrative suggests that communities attempt to transition to a relocalized world while it is still in their ability to do so, before the systems that support mobility collapse. Here, change inheres in the relationship between the self and the community, with the scope of change being so radical that current expertise and state authority become obsolete.

It is tempting at this point to adjudicate among the narratives – to assess, for instance, which is most likely to unfold, or which would result in a just or sustainable society.[16] It is tempting, too, to take a position on which form of subjecthood might be most fulfilling: whether it is better to move through the world as a driver-consumer, a neighborhood inhabitant, an adaptable human, or to become a *bricoleur* of the three, embracing different roles in different moments. Yet, the goal here is not to select a winner among the narratives. Ultimately, the analysis of these narratives does not necessarily tell us anything about the future; it can only tell us about the present. And in the present, these three narratives reveal an altogether fragmented American imagination. Policy discourse in the US is not progressing to a more comprehensive understanding of automobility (*contra* Cohen 2006). This is not a moment where policy-makers and the public have a coherent debate, weighing potential options and collectively choosing to 'accelerate,' or 'reverse,' or 'find the off ramp.'[17] Rather, the automobile is becoming enmeshed in a kaleidoscopic proliferation of stories that Americans tell about themselves.

This suggests a fragmentation of the dominance of automobility, even within one of its traditional bulwarks, the US Government. The authoritative representations of the future discussed in this paper make a hegemonic discursive move: they claim a situated world view as universal, and thus the imagined future for the urban US elite[18] becomes the official future of all. Yet, given the divergences among the three narratives, this is not hegemonic discursive closure – all texts converging on one urban-elite-colored view of the future – but instead what might be considered a fragmentation of hegemony.

Where does one go from here? What research agenda might emerge from the concerns evoked in this paper? I would suggest two interrelated approaches, both of which center around the fact that automobility – in its production, consumption, and consequences – is a global phenomenon. One approach would explore the broader transnational flow of narratives about the future of automobility. The narratives evoked in this paper are far from uniquely American. Smart growth, for instance, has proponents all over the world; it has grown out of an intellectual tradition of urban planning that has been thoroughly international since its inception. Mapping out the historical scope and global reach of these (and other) narratives about the future of the automobile would allow mobilities researchers to situate local instances of debate within a broader story of globally diffused imaginaries. It would highlight the ways in which local built environments are (and have been) shaped by transnationally mobile ideas. In so doing, it would perhaps provide new insights on the co-production of everyday mobilities and global politics.

This approach can be complemented by analysis of how narratives of the future play out in concrete instances across the globe (e.g. debates over specific transport projects). Given the dominance of the automobile in everyday urban life throughout much of the world, studying automobility is in large part about the nature of dominance and resistance. The basic analytic concept introduced in this paper – that 'futured' discourse shapes action and legitimates authority – gives us a new perspective on this dominance/resistance dynamic. The nominal future is a site where automobility's dominance is being de-routinized, problematized, and fragmented. Tracing out this discursive resistance as it plays out in a variety of local instances would offer up a perspective on the prevalence and depth of 'cracks' in the automobility regime. Such work would shed light not (merely) on how automobility might change in the future, but on how the world may already be witnessing a shift in the contours of mobility, authority, and identity.

Disclosure statement

No potential conflict of interest was reported by the author.

Notes

1. Throughout the paper, I take the term 'we' to mean 'we as mobile subjects in the twenty-first century.'
2. In North American parlance, an off ramp is an exit from a controlled-access highway, generally onto smaller and lower speed roads.
3. European city-dwellers drove an average of 5026 vehicle kilometers traveled (VKT) per capita annually in 1990; in Canada, this number was 7761; in Australia, 8034; and in the US, 12,336 (Kenworthy and Laube 1999).
4. Some take this history of critique as evidence that automobility has never been entrenched; its dominance has always been tenuous and contingent (Paterson 2007; Goodwin 2010; Dowling and Simpson 2013).
5. I come to discourse analysis largely via the literature on environmental political discourse (Litfin 1994; Hajer 1995; Epstein 2008). Hajer (1995) draws attention to the important political work that story-lines do in environmental policy-making. Epstein, in turn, draws on Laclau and Mouffe (1985) and Weldes (1999) to analyze subject positions within narratives.
6. Though automobility encompasses much more than simply the car itself, when building the sets of texts, I looked only for those texts specifically envisioning a future transformation of the automobile (whether this involved changes to the car itself or changes to the role the car plays in society).

7. The DOE has been the government agency responsible for energy-efficient vehicles since 1975, when the EPA transferred its advanced vehicle program to the forerunner of the DOE (Buck 1982).
8. The smart growth movement issued from a long history of social critique of suburbia, gaining institutional footing in the late 1990s, when the EPA formed the Smart Growth Network with several civil society partners, the White House launched a 'Livability Agenda,' and the Federal Transit Administration (FTA) published guidelines on building livable communities (FTA 1999). In this way, the anti-sprawl 'counter-public' in the last 20 years has increasingly begun to 'infiltrate regime practices, networks and discourses' (Sheller 2012, 184).
9. The Transition Movement, begun in the UK in the early 2000s, is an environmental and social movement that is rapidly expanding around the world. Transition US was launched in 2008.
10. These include the 'energy descent action plans' of various US cities (e.g. Berkeley, Bloomington, and Portland), which lay out policy proposals for their respective cities to transition to a post-peak-oil world.
11. These texts may also point to a discursive maneuver by 'carbon capital' (Urry 2013): several auto manufacturers and oil companies fund advanced vehicle R&D, perhaps as a means to adapt to a public discourse of climate change without abandoning the (lucrative) claim that modern society needs cars and roads.
12. These include the DOT and the EPA; their civil society partners include the US Green Building Council (USGBC), American Planning Association, and the Smart Growth Network among others.
13. The concept of smart growth shares much with the concepts of new urbanism and 'livability.' In general, 'livability' is broader than smart growth, while new urbanism places more emphasis on 'traditional' neighborhood form. See Ye, Mandpe, and Meyer (2005) for an analysis of different conceptualizations of 'smart growth.'
14. See Kumar (1991) on Arcadian utopian thought and Smith (2000) on the politics of 'nostalgia.'
15. See Garforth 2005 on ecological apocalyptic thought and Dryzek (2013) on survivalist discourse.
16. Indeed, weighing the ethics of future mobilities is a promising direction for further research and theory. There is much (and perhaps more systematic) work to be done in attempting to make sense of the play between action, imagination, and ethics (as in Jensen and Freudendal-Pedersen's [2012] work on utopian mobilities).
17. This suggests that some writings on risk society (e.g. Beck 1999) may be overly optimistic in their teleological rationalism: acknowledging modernity's self-inflicted risks does not necessarily lead to coherent, reasoned debate.
18. Such a future reflects classist or city-centric assumptions, including, among others, that people live close enough to the grid to be able to use an electric vehicle; that they do not have to transport heavy equipment on their commutes; or that they have leisure time in which to learn to grow food. Many thanks to an anonymous reviewer for this point.

References

Adey, P. 2006. "If Mobility is Everything Then It is Nothing: Towards a Relational Politics of (Im)Mobilities." *Mobilities* 1 (Mar): 75–94.

America's Energy Future Energy Efficiency Technologies Subcommittee, National Academy of Sciences, National Academy of Engineering, and National Research Council [AEF Panel et al.]. 2010. *Real Prospects for Energy Efficiency in the United States*. Washington, DC: National Academies Press. http://www.nap.edu/catalog.php?record_id=12621.

American Association of State Highway and Transportation Officials [AASHTO]. 2004. *A Policy on Geometric Design of Highways and Streets*. Washington, DC: AASHTO.

American Planning Association [APA]. 2006. *Planning and Urban Design Standards*. Hoboken, NJ: Wiley.

Anderson, B. R. 1991. *Imagined Communities: Reflections on the Origin and Spread of Nationalism*. New York: Verso.

Bærenholdt, J. O. 2013. "Governmobility: The Powers of Mobility." *Mobilities* 8 (Feb): 20–34.

Beck, U. 1999. *World Risk Society*. Malden, MA: Polity Press.

Beckmann, J. 2001. "Automobility – A Social Problem and Theoretical Concept." *Environment and Planning D: Society and Space* 19: 593–607.

Beckmann, J.. 2004. "Mobility and Safety." *Theory, Culture & Society* 21 (Oct 1): 81–100.

Berkeley Oil Independence Task Force. 2009. *Berkeley Energy Descent 2009–2020: Transitioning to the Post Carbon Era*. Berkeley, CA. http://postcarboncities.net/files/BerkeleyEnergyDescentPlan.pdf.

Bloomington Peak Oil Task Force. 2009. "Redefining Prosperity: Energy Descent and Community Resilience." http://bloomington.in.gov/peakoil.

Bohm, S., C. Jones, C. Land, and M. Paterson. 2006. "Introduction: Impossibilities of Automobility." In *Against Automobility*, edited by S. Bohm, C. Jones, C. Land, and M. Paterson, 3–16. Malden, MA: Blackwell.

Bonham, J. 2006. "Transport: Disciplining the Body That Travels." In *Against Automobility*, edited by S. Bohm, C. Jones, C. Land, and M. Paterson, 57–74. Malden, MA: Blackwell.

Buck, A. 1982. *A History of the Energy Research and Development Administration*. Washington, DC. http://energy.gov/sites/prod/files/ERDA%20History.pdf.

Bull, M. 2004. "Automobility and the Power of Sound." *Theory, Culture & Society* 21 (Oct 1): 243–259.

Chamberlin, S. 2009. *The Transition Timeline for a Local, Resilient Future*. White River Junction, VT: Chelsea Green.

Clarsen, G., and L. Veracini. 2012. "Settler Colonial Automobilities: A Distinct Constellation of Automobile Cultures?" *History Compass* 10 (Dec 2): 889–900.

Cohen, M. J. 2006. "A Social Problems Framework for the Critical Appraisal of Automobility and Sustainable Systems Innovation." *Mobilities* 1 (Mar): 23–38.

Collin-Lange, V. 2013. "Socialities in Motion: Automobility and Car Cruising in Iceland." *Mobilities* 8 (Sep): 406–423.

Conley, J., and A. T. McLaren. 2009. *Car Troubles*. Farnham: Ashgate.

Dalakoglou, D., and P. Harvey. 2012. "Roads and Anthropology: Ethnographic Perspectives on Space, Time and (Im)Mobility." *Mobilities* 7 (Nov): 459–465.

Dant, T. 2004. "The Driver-Car." *Theory, Culture & Society* 21 (Oct 1): 61–79.

Dennis, K., and J. Urry. 2009. *After the Car*. Cambridge: Polity.

Dowling, R., and C. Simpson. 2013. "'Shift – The Way You Move': Reconstituting Automobility." *Continuum* 27 (Jun): 421–433.

Dryzek, J. S. 2013. *The Politics of the Earth: Environmental Discourses*. Oxford: Oxford University Press.

Duany, A., J. Speck, and M. Lydon. 2010. *The Smart Growth Manual*. New York: McGraw-Hill.

Emirbayer, M., and A. Mische. 1998. "What is Agency?" *American Journal of Sociology* 103 (4): 962–1023.

Epstein, C. 2008. *The Power of Words in International Relations*. Cambridge, MA: MIT Press.

Featherstone, M. 2004. "Automobilities: An Introduction." *Theory, Culture & Society* 21 (Oct 1): 1–24.

Federal Transit Administration [FTA]. 1999. *Building Livable Communities with Transit*. Washington, DC: Federal Transit Administration.

Flink, J. J. 1988. *The Automobile Age*. Cambridge, MA: MIT Press.

Flonneau, M. 2010. "Read Tocqueville, or Drive? A European Perspective on US 'Automobilization'." *History and Technology* 26 (4) (Dec): 379–388.

Freund, P. E. S., and G. T. Martin. 1993. *The Ecology of the Automobile*. New York: Black Rose Books.

Garforth, L. 2005. "Green Utopias: Beyond Apocalypse, Progress, and Pastoral." *Utopian Studies* 16 (3): 393–427.

Gartman, D. 2004. "Three Ages of the Automobile the Cultural Logics of the Car." *Theory, Culture & Society* 21 (Oct 1): 169–195.

Geels, F. W. 2012. *Automobility in Transition?: A Socio-technical Analysis of Sustainable Transport*. New York: Routledge.

Gilbert, R., and A. Perl. 2010. "Transportation in the Post-Carbon World." In *The Post Carbon Reader*, edited by R. Heinberg and D. Lerch, 347–362. Berkeley, CA: Watershed Media.

Goddard, S. B. 1994. *Getting There: The Epic Struggle between Road and Rail in the American Century*. New York: Basic Books.

Goodwin, K. J. 2010. "Reconstructing Automobility: The Making and Breaking of Modern Transportation." *Global Environmental Politics* 10 (4): 54–58.

Goodwin, P., and K. Van Dender. 2013. "Peak Car – Themes and Issues." *Transport Reviews* 33 (Jun 20): 243–254.

Gunster, S. 2004. "'You Belong Outside': Advertising, Nature, and the SUV." *Ethics & the Environment* 9 (2): 4–32.

Hagman, O. 2006. "Morning Queues and Parking Problems. On the Broken Promises of the Automobile." *Mobilities* 1 (Mar): 63–74.
Hajer, M. A. 1995. *The Politics of Environmental Discourse*. Oxford: Clarendon Press.
Heinberg, R. 2003. *The Party's Over: Oil, War and the Fate of Industrial Societies*. Gabriola, BC: New Society Publishers.
Hopkins, R. 2008. *The Transition Handbook: From Oil Dependency to Local Resilience*. Totnes: Green.
Hopkins, R. 2011. *The Transition Companion: Making Your Community More Resilient in Uncertain times*. White River Junction, VT: Chelsea Green.
Huijbens, E. H., and K. Benediktsson. 2007. "Practising Highland Heterotopias: Automobility in the Interior of Iceland." *Mobilities* 2 (Mar): 143–165.
Jackson, K. T. 1985. *Crabgrass Frontier: The Suburbanization of America*. New York: Oxford University Press.
Jasanoff, S. 2004. *States of Knowledge*. London: Routledge.
Jensen, A. 2011. "Mobility, Space and Power: On the Multiplicities of Seeing Mobility." *Mobilities* 6 (May): 255–271.
Jensen, O. B., and M. Freudendal-Pedersen. 2012. "Utopias of Mobilities." In *Utopia: Social Theory and the Future*, edited by M. Hviid Jacobsen and K. Tester, 197–218. Farnham: Ashgate.
Joas, H. 1996. *The Creativity of Action*. Chicago: University of Chicago Press.
Kent, J. L., and R. Dowling. 2013. "Puncturing Automobility? Carsharing Practices." *Journal of Transport Geography* 32 (Oct): 86–92.
Kenworthy, J. R., and F. B. Laube. 1999. "Patterns of Automobile Dependence in Cities." *Transportation Research Part A: Policy and Practice* 33: 691–723.
Koonin, S. E. 2011. "Statement of Steven E. Koonin, under Secretary for Science." US Department of Energy, before the Committee on Energy and Natural Resources, United States Senate. November 15. Washington, DC.
Koshar, R. 2004. "Cars and Nations Anglo-German Perspectives on Automobility between the World Wars." *Theory, Culture & Society* 21 (Oct 1): 121–144.
Kumar, K. 1991. *Utopianism*. Minneapolis, MN: University of Minnesota Press.
Kunstler, J. H. 1994. *The Geography of Nowhere: The Rise and Decline of America's Man-Made Landscape*. New York: Simon & Schuster.
Laclau, E., and C. Mouffe. 1985. *Hegemony and Socialist Strategy: Towards a Radical Democratic Politics*. London: Verso.
Latimer, J., and R. Munro. 2006. "Driving the Social." *The Sociological Review* 54 (Sep 18): 32–53.
Litfin, K. 1994. *Ozone Discourses*. New York: Columbia University Press.
Lumsden, K. 2013. "(Re)Civilizing the Young Driver: Technization and Emotive Automobility." *Mobilities* 10 (Aug 15): 36–54.
MacIntyre, A. C. 1984. *After Virtue: A Study in Moral Theory*. Notre Dame: University of Notre Dame Press.
McLaren, A. T., and S. Parusel. 2012. "Under the Radar: Parental Traffic Safeguarding and Automobility." *Mobilities* 7 (May): 211–232.
Merriman, P. 2004. "Driving Places: Marc Augé, Non-Places, and the Geographies of England's M1 Motorway." *Theory, Culture & Society* 21 (Oct 1): 145–167.
Merriman, P. 2006. "'Mirror, Signal, Manoeuvre': Assembling and Governing the Motorway Driver in Late 1950s Britain." *The Sociological Review* 54 (Sep 18): 75–92.
Metz, D. 2014. *Peak Car*. London: Landor LINKS.
Miller, D. 2001. *Car Cultures*. Oxford: Berg.
Milliken, J. 1999. "The Study of Discourse in International Relations: A Critique of Research and Methods." *European Journal of International Relations* 5 (Jun 1): 225–254.
Mumford, L. 1961. *The City in History*. New York: Harcourt, Brace & World.
Nader, R. 1965. *Unsafe at Any Speed*. New York: Grossman.
National Research Council [NRC]. 2013. *Review of the Research Program of the U.S. DRIVE Partnership: Fourth Report*. Washington, DC: National Academies Press.
Oak Ridge National Laboratory. 2010. *Report on the Transportation Combustion Engine Efficiency Colloquium Held at USCAR, March 3–4, 2010*. Oak Ridge, TN: Oak Ridge National Laboratory.
Packer, J. 2008. *Mobility without Mayhem*. Durham, NC: Duke University Press.
Paterson, M. 2007. *Automobile Politics*. Cambridge: Cambridge University Press.
Pearce, L. 2012. "Automobility in Manchester Fiction." *Mobilities* 7 (Feb): 93–113.

Portland Peak Oil Task Force. 2007. "Descending the Oil Peak: Navigating the Transition from Oil and Natural Gas." http://www.portlandonline.com/shared/cfm/image.cfm?id=145732.
President's Council of Advisors on Science and Technology [PCAST]. 2010. "Report to the President on Accelerating the Pace of Change in Energy Technologies through an Integrated Federal Energy Policy." http://www.whitehouse.gov/sites/default/files/microsites/ostp/pcast-energy-tech-report.pdf.
Rajan, S. C. 2006. "Automobility and the Liberal Disposition." *The Sociological Review* 54 (Sep 18): 113–129.
Reese, K. 2015. "The Traffic Light." In *Making Things International 1: Circuits and Motion*, edited by M. Salter, 36–48. Minneapolis, MN: University of Minnesota Press.
Ricoeur, P. 1991. *From Text to Action*. Evanston, IL: Northwestern University Press.
Sachs, W. 1992. *For Love of the Automobile: Looking Back into the History of Our Desires*. Berkeley: University of California Press.
San Francisco Peak Oil Preparedness Task Force. 2009. *Report of the San Francisco Peak Oil Preparedness Task Force*. San Francisco: City of San Francisco. http://postcarboncities.net/node/4374.
Scott, J. C. 1998. *Seeing like a State: How Certain Schemes to Improve the Human Condition Have Failed*. New Haven, CT: Yale University Press.
Seiler, C. 2008. *Republic of Drivers*. Chicago, IL: University of Chicago Press.
Seiler, C. 2010. "Author Response: The Ends of Automobility." *History and Technology* 26 (Dec): 389–397.
Sheller, M. 2004. "Automotive Emotions: Feeling the Car." *Theory, Culture & Society* 21 (Oct 1): 221–242.
Sheller, M. 2012. "The Emergence of New Cultures of Mobility: Stability, Openings, and Prospects." In *Automobility in Transition? A Socio-technical Analysis of Sustainable Transport*, edited by F. W. Geels, R. Kemp, G. Dudley, and G. Lyons, 180–202. New York: Routledge.
Sheller, M., and J. Urry. 2000. "The City and the Car." *International Journal of Urban and Regional Research* 24 (4): 737–757.
Smart Growth Network, International City/County Management Association, and US Environmental Protection Agency. 2002. *Getting to Smart Growth: 100 Policies for Implementation. Development*. Washington, DC: Smart Growth Network. http://www.smartgrowth.org/pdf/gettosg.pdf.
Smith, K. K. 2000. "Mere Nostalgia: Notes on a Progressive Paratheory." *Rhetoric & Public Affairs* 3 (4): 505–527.
Transition United States [Transition US]. 2011. "Transition Primer: A Guide to Becoming a Transition Town." http://transitionus.org/sites/default/files/primer-us-v2.0.pdf.
Transition United States [Transition US]. 2012. "A Year in Review." http://transitionus.org/sites/default/files/TUS-2012-AYearinReview.pdf.
Transition United States [Transition US]. n.d. "Governance Toolbox V2.0." http://transitionus.org/guides/governance-toolkit.
Urry, J. 2004. "The 'System'of Automobility." *Theory, Culture & Society* 21 (Oct 1): 25–39.
Urry, J. 2006. "Inhabiting the Car." *The Sociological Review* 54 (s1): 17–31.
Urry, J. 2007. *Mobilities*. Cambridge: Polity.
Urry, J. 2013. *Societies Beyond Oil: Oil Dregs and Social Futures*. London: Zed Books.
US Department of Energy [US DOE]. 2006. *Basic Research Needs for Clean and Efficient Combustion of 21st Century Transportation Fuels*. Washington, DC: US Department of Energy. http://www.osti.gov/servlets/purl/935428-bbBji1/.
US Department of Energy [US DOE]. 2009. *Advanced Combustion Engine R&D: Goals, Strategies, and Top Accomplishments*. Washington, DC: US Department of Energy
US Department of Energy [US DOE]. 2010. *Multi-Year Program Plan*. Washington, DC: US Department of Energy.
US Department of Energy [US DOE]. 2011. *Report on the First Quadrennial Technology Review*. Washington, DC: US Department of Energy.
US Department of Energy [US DOE]. 2012. *Report on the First Quadrennial Technology Review: Technology Assessments*. Washington, DC: US Department of Energy.
US Department of Transportation [US DOT]. 2010. *Livability in Transportation Guidebook*. Washington, DC. http://www.fhwa.dot.gov/livability/case_studies/guidebook/livabilitygb10.pdf.
US Environmental Protection Agency [US EPA]. 2009. "Essential Smart Growth Fixes for Urban and Suburban Zoning Codes." Washington, DC. http://www.epa.gov/dced/pdf/2009_essential_fixes.pdf.

US Green Building Council [USGBC]. 2009. *Green Neighbourhood Development: LEED Reference Guide for Neighbourhood Development*. Washington, DC: US Green Building Council. http://www.usgbc.org/resources/leed-neighbourhood-development-v2009-current-version.

Watson, M. 2012. "How Theories of Practice Can Inform Transition to a Decarbonised Transport System." *Journal of Transport Geography* 24 (Sep): 488–496.

Weldes, J. 1999. *Constructing National Interests: The United States and the Cuban Missile Crisis*. Minneapolis: University of Minnesota Press.

Wollen, P., and J. Kerr. 2002. *Autopia: Cars and Culture*. London: Reaktion Books.

World Bank. 2003. *World Development Indicators 2003*. Washington, DC: World Bank.

Ye, L., S. Mandpe, and P. B. Meyer. 2005. "What is 'Smart Growth?' – Really?" *Journal of Planning Literature* 19 (Feb 1): 301–315.

Index

Note: **Boldface** page number refer to figures and tables, page number followed by n denote endnotes.

Adey, Peter 18, 85
ADS-B *see* Automatic Dependent Surveillance-Broadcast
Aerial Life (Adey) 18
aeromobilities 17–20; overview of 49–50; 'Pinning Down' 51–2; re-assembling of 52–3, 61; research 52; in Singapore *see* Singapore's aeromobilities; territorialisations 61
Africa, disease mobilities 23–4
aircraft monitoring systems 51
air-mindedness 137–8
air services agreements (ASAs) 55
air traffic services, in flight information regions 58–9
air transport system 51
Architectural Review, The (Richards) 88
ASAs *see* air services agreements
assemblage theory 52–3, 61
Automatic Dependent Surveillance-Broadcast (ADS-B) 57–8
automobility: 'accelerate' 156–9; change and continuity 155–6; in crisis 154–5; 'find the off ramp' 157, 162–4; in modernity 153–4; overview of 152–3; 'reverse' 156–7, 159–62

Ben-Joseph, Eran 94
brave encounters, timid encounters *vs.* 110

Carol's smartphone, map app on 123
car parking, in Britain: affective environments and economies 90–2; infrastructures and practices 92–4; modernist dreams and urban nightmares 88–90; and policies 95
Centre for Mobilities Research (CeMoRe) 1
chance encounters, pseudonymous strangers and 108–10
Chaos of Disciplines (Abbott) 67
citizen space exploration 140
'city of strangers' 106
civil aviation 54–6
civil inattention 101, 107, 113
collocated encounters, pseudonymous strangers and 107–8
Cowen, Deborah 22–3

De Landa, M. 52
digital mobilities, locative media and *see* locative media
direct environmental costs, of parking policies 92
disease mobilities: Cholera 24–5; Ebola virus 25; governance and peacekeepers 24–5; HIV infection 24; humanitarian responses 26–7; non-human mode of 23–4

Ebola outbreak 25
'ecosystem services' 23, 28
elite secession, geo-ecologies of: infrastructures 20; kinetic elites 22; logistics 22–3; network capital 21; offshoring 21–2
Elliott, Anthony 27
EMCA *see* ethnomethodology and conversations analysis
emergency mobilities: anticipation 36–7; assumptions in 35–6; coordination 38; distributional differences 43; inequalities emergencies in 34; inhuman speeds 41–2; mobile machines 38–9; necromobilities and geographies 40; overview of 32–4; social difference 42–3; 'state of emergency' 34; time limit 43–4
'energy descent' plans 162–3, 166n10
environmental parking policies 92–4
Environmental Protection Agency (EPA) 157, 166n8, 166n12
ethnomethodology and conversations analysis (EMCA) 119–20

Federal Transit Administration (FTA) 166n8
Fire Service Emergency Cover Toolkit 37
flight information regions (FIRs) 58–60, **59,** 62n1

Foucauldian approach 15–28
'Freedoms of the Air' 55
FTA *see* Federal Transit Administration
Fuller, Gillian 20

global assemblage 53, 57, 61
Global Complexity (Urry) 85
global shadows, Ferguson's notion of 16
Goffman's civil inattention 101, 107, 113
Gordon, Rodney 89
governmobility concept 16

Haiti, disease mobilities 24
Hannam, Kevin 34
Hooker, Charlie 89
hybrid ecology, seamfulness and 103–5

ICAO *see* International Civil Aviation Organisation
ICAP *see* international civil aviation policy
indirect environmental costs, of parking policies 92
Intelligent Transportation Systems 38
interdisciplinary research fields: complexity of 68; migrant and exceptionalism 74–6; of migration and mobilities 67–8; nations and methodological nationalism 73–4; sequencing of ideas 70–3; sometimes-migrants, cases of 76–9; transnationalism 69–70
International Civil Aviation Organisation (ICAO) 54; Asia-Pacific Regional Air Navigation Meeting in 1973 59; 'standards and recommendations' 60
international civil aviation policy (ICAP) 56
'islanding effect' 26

Katz, Jonathan 27
kinetic elites 22

Lavau, Stephanie 24
Lebeau, Andre 135
locative media: empirical basis and methods 102–3; hybrid ecology 104–5; overview of 100–2; pseudonymous strangers, encounters with *see* pseudonymous strangers; seams 103–4; 'unfolding' concept 105–6; user experience 105

Marples, Ernest 88
Mavhunga, C. C. 23–4
McCluskey, Jim 91
McFarlane, Colin 27–8
methodological nationalism 73–4, 137
migrant exceptionalism 6, 74–6
mining sector 25–6
mobility infrastructures/mobile infrastructuring: approaches to 85; elemental materialities of 86; Lefebvrian concept of 'moorings' 84–5; movements 86; overiew of 83–4; practices of 87
'mobility turn' 100
molecular mobilities 86
'monologic imagination' 137
moorings, Lefebvrian concept of 84
motility, definition of 20
Mumford, Lewis 88

network capital 21
new mobilities paradigm 4, 16
non-human mobilities 23–4

Osterholm, Michael 26

Parr, Martin 90
pedestrian mobility: 'doing walking' 119–20; leader-follower arrangement 129–31; lightweight video cameras 122; mediated walking 120–1; overview of 117–18; reconnaissance and re-stopping 126–9; unilateral stopping 124–5; walking and checking route 122–4; you-are-here dot 125–6
personal spaceflight, definition of 140–1
'post-carbon' world 162–3
pseudonymous strangers: chance encounters 108–10; collocated encounters 107–8; concept of 101–2, 106–7; Goffman's civil inattention 101, 107, 113; timid encounters 110–13

racialized space, tourist mobilities and 17–20
reflexive hybrid ecologies 105
relative immobilities 85
Richards, J. M. 88

seamfulness, and hybrid ecology 103–5
secessionist mobility 21
Securing Paradise: Tourism and Militarism in Hawai'i and the Philippines (Vicuña Gonzalez)
Sheller, Mimi 34
SIA *see* Singapore Airlines
Singapore Airlines (SIA) 56
Singapore's aeromobilities: aeromobile territorialisations 61; airline-and-air-hub model 56; flight information regions 58–60, **59**, 62n1; on geopolitics 58–60; intensified airspace 57; on route regulation 54–6; on technology 56–8
smart growth 159–60, 165, 166n8, 166n13
South China Sea 57

space tourism: aeromobilities and tourism mobilities 137–9; anticipation and prediction 139–41; comparisons with existing mobilities 142–3; competition to 145–6; determinism and 146–7; integration with society 143–5; overview of 135–6

technogeopolitics 58
ten years of mobilities research: maturation and challenges 8–11; overview of 1–3; politics of 7–8; special issue papers 5–7; trajectories of 3–5
timid encounters: with pseudonymous strangers 110–13; *vs.* brave encounters 110
tourist mobilities 17–20
'traditional' transport geography 51
Transition Movement 157, 166n9
Transition US 157, 162–4
transnationalism, asynchronous engagement with 69–70
Trinity Square 89
22nd Assembly in 1977 60

uneven mobility: disease outbreaks and 23–7; elite secession, geo-ecologies of 20–3; overview of 15–17, 27–8; tourism and 17–20
'unfolding' concept 105–6
Urry, John 21, 34, 85
US Department of Energy (US DOE) 156–9, 166n7
US Department of Transportation (US DOT) 157, 160–2, 166n12
US Green Building Council (USGBC) 160

Vasudevan, Alex 27–8
Vicuña Gonzalez, Vernadette 19

Westphalian system, of flight information regions 60

For Product Safety Concerns and Information please contact our EU representative GPSR@taylorandfrancis.com Taylor & Francis Verlag GmbH, Kaufingerstraße 24, 80331 München, Germany

Batch number: 10732736

Printed by Printforce, the Netherlands